Y0-CAS-674

Centennial History of the Carnegie Institution of Washington
Volume III   The Geophysical Laboratory

For over a century, the Geophysical Laboratory of the Carnegie Institution of Washington has witnessed exciting discoveries and ingenious research, made possible by the scientific freedom granted to members of the department. For the most part, this research has involved laboratory experimentation on the physics and chemistry of rock-forming minerals at high temperature and pressure. This third volume in a series of five histories of the Carnegie Institution documents the contribution made by the members of the Geophysical Laboratory to our understanding of the Earth, from mineral formation deep below the surface, to the search for the origins of life, and out into space to study the chemical evolution of the interstellar medium. Field work has taken researchers from active volcanoes to ships collecting ocean sediments, and geological mapping expeditions around the world. Contemporary photographs throughout illustrate the evolution of the department and its research.

HATTEN S. YODER, JR., internationally known experimental petrologist and geochemist, and Emeritus Director of the Geophysical Laboratory, died in August 2003. After completion of his Ph.D. at the Massachusetts Institute of Technology in 1948, he spent his scientific career at the Geophysical Laboratory. He was also visiting professor at CalTech, University of Texas, University of Colorado, and University of Cape Town, and held honorary doctorate degrees from the University of Paris VI and the Colorado School of Mines. In addition to his services in numerous professional societies, he was a member of the National Research Council's Executive Committee, as well as the US National Committees for Geochemistry, and History of Geology. He advised US Congress on issues ranging from natural resources to the hazards of asbestos.

CENTENNIAL HISTORY OF THE CARNEGIE INSTITUTION OF WASHINGTON

Volume III

THE GEOPHYSICAL

LABORATORY

HATTEN S. YODER, Jr.

Formaly Director Emeritus,
Geophysical Laboratory

CAMBRIDGE
UNIVERSITY PRESS

PUBLISHED BY THE PRESS SYNDICATE OF THE UNIVERSITY OF CAMBRIDGE
The Pitt Building, Trumpington Street, Cambridge, United Kingdom

CAMBRIDGE UNIVERSITY PRESS
The Edinburgh Building, Cambridge CB2 2RU, UK
40 West 20th Street, New York, NY 10011–4211, USA
477 Williamstown Road, Port Melbourne, VIC 3207, Australia
Ruiz de Alarcón 13, 28014 Madrid, Spain
Dock House, The Waterfront, Cape Town 8001, South Africa

http://www.cambridge.org

© Carnegie Institution of Washington 2004

This book is in copyright. Subject to statutory exception
and to the provisions of relevant collective licensing agreements,
no reproduction of any part may take place without
the written permission of Cambridge University Press.

First published 2004

Printed in the United Kingdom at the University Press, Cambridge

Typeface Adobe Garamond 10.75/12.5 pt.   *System* LATEX 2ε   [TB]

A catalogue record for this book is available from the British Library

ISBN 0 521 83080 X hardback

This history of the Geophysical Laboratory was completed by Hatten S. Yoder, Jr. just a few weeks before his death at age 82. Hat was an exceptional scientist, a valued colleague and a fine gentleman. He served his country and his science community well, and was awarded many honors by his peers. A member of the National Academy of Science, he was an aficionado of art, science, and history who kept an open mind on controversial subjects. The Carnegie Institution and the Geophysical Laboratory are proud and grateful that he was one of us. We dedicate this volume to him.

Wesley T. Huntress, Jr., Director of the Geophysical Laboratory

CONTENTS

Contents

FOREWORD

In 1902 Andrew Carnegie, a steel magnate turned philanthropist, had a brilliant idea. Carnegie was prescient in recognizing the important role that science could play in the advancement of humankind. He also believed that the best science came by providing "exceptional" individuals with the resources they need in an environment that is free of needless constraints. He created the Carnegie Institution as a means to realize these understandings, directing the Institution to undertake "projects of broad scope that may lead to the discovery and utilization of new forces for the benefit of man." Carnegie was confident that this unusual formula would succeed. And he was right.

For over a century, the Carnegie Institution has sponsored creative and often high-risk science. Some of the luminaries who were supported by the Institution over the years are well known. For example, Edwin Hubble, who made the astonishing discoveries that the universe is larger than just our galaxy and that it is expanding, was a Carnegie astronomer. Barbara McClintock, who discovered the existence of transposable genes, and Alfred Hershey, who proved that DNA holds the genetic code, both won Nobel Prizes for their work as Carnegie scientists. But many other innovative Carnegie researchers who are perhaps not so well known outside their fields of work have made significant advances.

Thus, as part of its centennial celebration, the Institution enlisted the help of many individuals who have contributed to the Institution's history to chronicle the achievements of the Institution's five major departments. (Our newest department, the Department of Global Ecology, was started in 2002 and its contributions will largely lie ahead.) The result is five illustrated volumes, which describe the people and events, and the challenges and controversies behind some of the Institution's significant accomplishments. The result is a rich and fascinating history not only of the Institution, but also of the progress of science through a remarkable period of scientific discovery.

Andrew Carnegie could not have imagined what his Institution would accomplish in the century after its founding. But I believe that he would be very proud. His idea has been validated by the scientific excellence of the exceptional men and women who have carried out his mission. Their work has placed the Institution in a unique position in the world of science, which is just what Andrew Carnegie set out to do.

RICHARD A. MESERVE
President, Carnegie Institution of Washington

PREFACE

In celebration of the Centennial of the Carnegie Institution of Washington, the then President Maxine F. Singer requested that each department prepare a history of its scientific accomplishments. I was invited to expand my brief history, "Scientific highlights of the Geophysical Laboratory, 1905–1989," which I had prepared on the occasion of the move to a new building on the campus of the Department of Terrestrial Magnetism in 1990. As with most histories, it records the biased views of a single observer moderated by the available written records and evaluations of others. The early records of the Geophysical Laboratory are minuscule, and even though I have personally witnessed over half the centennial (55 years), memories are not always reliable. It is difficult to subdue the enthusiasm, admiration and pride I have for the Geophysical Laboratory and its past and present members. Having known almost all of the early staff members, perhaps I might be forgiven for any excessive claims of discovery attributed to the staff. All science is built on the discoveries of others, and it is not always evident who arrived at the pinnacle of an idea first, demonstrated its proof, and applied the idea to the solution of a geological problem or capitalized on its promotion. The personal satisfaction of contributing to the growth of science is adequate reward in itself.

The most important factor in the generation of new ideas at the Geophysical Laboratory has been the freedom of choice to follow whatever a staff member believes to be important in the solution of a geological problem. The scientist's overriding goal was to achieve an understanding of the problem so that the critical variables could be recognized, evaluated, and formulated into general concepts useful in solving other problems. The intent, therefore, was to seek knowledge that has broad application to the major problems of the Earth. Because no researcher can predict how a fundamental discovery might be applied to future societal needs or problems, there is no test for relevance or applicability applied to the work at the Geophysical Laboratory as is made in industrial organizations. That freedom to follow whatever is critical to the solution of problems is why the Geophysical Laboratory has remained unique among research organizations. The price of such a generous measure of scientific freedom is greater personal responsibility to produce and greater accountability. Although peer review provides for continual testing, the responsibility to produce was self-generated and was expressed by the high motivation and involvement of the staff.

In this regard, the Director has had the responsibility to select staff who exhibited the breadth of understanding, dedication, and enthusiasm for the projects they selected. On a few rare occasions an assistant was hired to carry out certain measurements, but in general, each staff member did his or her own work. No one was treated as a "pair of hands" and everyone was expected to contribute to the generation of the science itself. Collaboration was encouraged and took place by mutual agreement where special talents could contribute. Over time as the science became more interdisciplinary, the amount of cooperation has indeed increased.

The range of projects covered in the following chapters is great, but unfortunately, not complete. The extensive documentation was intended to achieve accuracy and to encourage the reader to seek out the details of the problem investigated. I do not pretend to be an expert in all the areas covered, and, hopefully, will be forgiven for errors, and omissions. Perhaps the excitement of discovery and ingenuity of research will retain the reader's attention.

ACKNOWLEDGMENTS

The cooperation of the staff of both the Geophysical Laboratory and the Department of Terrestrial Magnetism has been particularly helpful in reviewing the chapters close to their interests. Of great assistance in the preparation of the chapters was the availability of the "Indices of the Annual Reports of the Director of the Geophysical Laboratory, 1905–1980," prepared by Robert M. Hazen and Mrs. Margaret H. Hazen and issued as separate paper No. 1860 of the Geophysical Laboratory in 1981. Although I hold an extensive collection of the books and reprints of papers published by the early staff members, it was the timely service of librarians Shaun Hardy and Merri Wolf, who borrowed a substantial number of books and papers from libraries throughout the country, that provided much of the background reading material. Most appreciated was the kind patience of Mrs. Susan A. Schmidt who typed the manuscript and made the many insertions, corrections, and format changes required. All are thanked for their help in providing an accurate history of the scientific work of the Geophysical Laboratory, which will in turn be celebrating its own centennial in 2005.

IMPERIAL TO METRIC
CONVERSION FACTORS

1 inch = 2.54 cm
1 foot = 30.48 cm
1 yard = 91.44 cm
1 cubic inch = 16.387 cubic cm
1 cubic foot = 0.0283 cubic m
1 acre = 0.4047 ha
1 square mile = 2.5899 square km
1 mile = 1.609 km
1 nautical mile = 1.852 km
1 pound = 0.4536 kg

INTRODUCTION

Recognition of "geophysics"

The term for the physical investigation of the Earth was first defined by Fröbel in 1834.[1] He wrote:

> One can say geophysics, if one means such conditions of the earth to which the various chapters of experimental physics correspond, thereby excluding geographical organization of organic nature as well as the geographical distribution of minerals.

He realized that at that time "much lies at hand to be designated by this name, indeed only the idea exists, and in addition I will possibly suppose, that a tendency lies in what is currently called geology to expand it to mean the theory of the Earth."[2] In succeeding years, the fields of geomagnetism, seismology, geodesy, and meteorology were included under the term.

Initiation of geophysics in the USA

The work in experimental geophysics was initiated in the USA by Clarence King, appointed as the first director of the US Geological Survey (USGS) in 1879. His two-fold program involving the study of mining districts and the collection of mineral statistics was supplemented by geophysical and geochemical studies in support of mining geology.[3] In 1880, Carl Barus, a physicist, was assigned to George F. Becker, a mining geologist with broad interests in physical and chemical problems, to make an experimental determination of the electrical currents around ore bodies. With Strouhal in 1885, he found a relation between electrical conductivity and the temperature of steels in different states of hardness, wrought iron, and cast iron, whose differences were being debated at the time.[4]

Emphasis on geological mapping

King resigned in 1881 and was replaced by John W. Powell, who emphasized geological mapping, and the experimental work in geophysics declined. Nevertheless, in 1884 the USGS moved into new headquarters and a physical laboratory as well as chemical laboratories were established. Following his mathematical bent, Becker considered the fundamental shape of volcanic cones and the mechanical conditions for faulting. In addition, he developed ideas on how chemical energy would influence the succession of minerals in volcanic rocks. Eventually he became deeply involved in the study of strain, flow, and rupture in rocks.

Physics and chemistry terminated at USGS

In 1892 there was a major reduction in funding for the USGS, and the appropriations for physics and chemistry terminated. Becker was dismissed but reinstated in 1894 when Charles D. Walcott became director of the USGS to carry out reconnaissance geology of the gold areas in the southern Appalachians. In the meantime, Barus had studied the electromotive force of various thermocouples, the melting temperature of rocks, and the volume change on melting of diabase. In spite of his success he was dismissed. Although no longer regarded as important at the USGS, geophysics attracted the attention of universities.

Faculty positions at universities

In 1896 appointments to individual faculty positions in geophysics were made at two universities: Louis A. Bauer in geomagnetism at the University of Chicago, and Harry F. Reid in dynamic geology and seismology at Johns Hopkins University. The first university chair in geophysics in Germany was founded at the Institute of Geophysics in Göttingen in 1898; however, work there on geophysical problems began as early as 1756. Geophysics was clearly a discipline that could no longer be ignored.

Physics and chemistry re-established at USGS

In 1900 the Division of Physical and Chemical Research was re-established in the USGS under Becker's direction, but the experimental work was carried out by two new employees, Arthur L. Day, an American staff worker at the Physikalische-Technische [Reichs] Anstalt in Berlin, and E. T. Allen, a chemist. They determined the melting temperature of the principal rock-forming minerals, beginning with the plagioclase feldspars. This research was indeed critical to an understanding of the physical chemistry of rocks.

Education vs. research

While the new work was progressing, a new opportunity emerged. The wife of the director of the USGS, Mrs. Walcott, was active in a group in Washington whose aim was to start a national university in honor of George Washington.[5] The Washington Memorial Institution was incorporated with Charles D.

Walcott as president of the Board of Trustees and Daniel C. Gilman as director. It was clear that substantial funding would be needed, so Walcott suggested that Gilman meet with Andrew Carnegie. The meeting in October 1901 included Mrs. Walcott, but they were unsuccessful in persuading Mr. Carnegie to support a national university. However, a plan for establishing a research institution in Washington was considered more appropriate. On November 16, 1901 Gilman and John S. Billings met with Carnegie and the emphasis was switched from education to research and postgraduate training.

Carnegie gift for research institution

Events moved quickly thereafter when Carnegie wrote to President Theodore Roosevelt on November 28, 1901 that he propose a gift to the nation. On December 2, 1901, Gilman, Billings, and Carnegie had lunch with President Roosevelt and a gift of $10 million was announced for a scientific research institution in Washington along the lines of a memorandum prepared by Walcott. (Carnegie added $2 million to the endowment in 1909 and an additional $10 million in 1911.) That afternoon Carnegie met with Walcott[6] at the New Willard Hotel[7] that had just opened in 1901 on Pennsylvania Avenue in Washington, and told him of his wishes that his old friend Gilman, first president of Johns Hopkins University, be president of the new institution and that Walcott take an active part as secretary in conducting the institution in light of Gilman's advanced age (he was seventy years old). Walcott agreed, taking note, however, that his primary duty was as director of the USGS.

The magnificent scheme

Early in December 1901, Walcott asked Becker to prepare quickly a statement for an independently endowed geophysical laboratory. Walcott's enthusiasm for such a laboratory may have stemmed from a talk given earlier that year by Arthur L. Day at the Philosophical Society of Washington on January 19, 1901. Walcott, as president, had heard Day's "account of the history, organization, and work of the Physikalische-Technische [Reichs] Anstalt of Berlin".[8]

On December 16, 1901, Becker delivered his outline for a geophysical laboratory to Walcott under the title "Concerning a Geophysical Laboratory." A handwritten draft with corrections and a typed version are preserved.[9] The first paragraph of the typescript is as follows:

> It is difficult to conceive of a more magnificent scheme than the founding of a generously endowed laboratory, devoted to researches into the physical and chemical conditions affecting the history of the globe. Very little work has been done in this direction for several reasons. As a rule physicists and chemists know too little of

geology to appreciate the applicability of these sciences to the elucidation of the history of the earth, while few geologists have the training in exact science which would fit them to undertake such researches independently. Again the investigations required are so laborious and expensive that no institution now in existence is in a position to undertake them, systematically, on an organized plan. The nearest approach to such work is that now going on in the division of Chemical and Physical Research of the Geological Survey, but there is little ground for hope of adequate appropriations from Congress for this purpose.

The two-page statement was presumably too brief in scientific substance for Walcott to use to persuade the Carnegie Institution of Washington (CIW) Board of Trustees, incorporated on January 4, 1902, for which he had been elected secretary, and a more detailed report was requested. Becker submitted the desired scientific program to Walcott on March 21, 1902.[10] The program emphasized the need for experiments on the physical properties of rocks as they applied to "terrestrial density, upheaval and subsidence, and volcanism." The principal projects were listed under (1) Mechanics, (2) High-temperature work, (3) Solutions and their relations, (4) Thermodynamics, and (5) Constitution of matter. The scientific program was published in 1903 with detailed plans for staff, building design, budget, and organization in an Appendix to the Report of the Advisory Committee on Geophysics.[11]

Exceptional man vs. central laboratories

In accord with Carnegie's wishes, the funds were to be used in "securing the exceptional man for the work for which he is intended, and supplying the necessary apparatus for experiments and research" (Letter from Carnegie to Donaldson, December 20, 1901). In contrast, the Executive Committee of the new Institution instituted the concept of small advisory committees to prepare reports on the needs of specific fields. The issue of individualism versus collectivism was clearly drawn. As Walcott explained to Carnegie, "individualism is the old view that one man can develop and carry forward any line of research, whereas collectivism embodies the modern idea of cooperation and community of effort." In the years to follow, independent research departments developed in which teams of investigators tackled major scientific problems. R. S. Woodward, second president of CIW, described it as "a university in which there are no students."[12]

Grants to individuals

Grants were given in 1903 to Frank D. Adams at McGill University (No. 4) to study the flow of rocks (Carrara marble, dolomites, and limestones) under pressure; and to C. R. Van Hise at the University of Wisconsin

(No. 71) to investigate the state of geophysical research in European institutions. In 1904 grants were given to George F. Becker of the USGS (No. 172) to study elasticity and plasticity on solids; to Arthur L. Day also in the USGS (No. 171) for the experimental investigation of mineral fusion and solution under pressure; to G. K. Gilbert (No. 126) for the preparation of plans to investigate the thermal gradient in a deep drill hole of at least 6000 feet depth; to Carlos de Mello (No. 170) for a bibliography on geophysics; as well as a continuation of the grant to F. D. Adams (No. 117) for studying the flow of rocks under pressure. The grants to Becker (No. 172) and Day at the USGS (No. 225); to F. D. Adams at McGill University (No. 227); and the bibliography for geophysics[13] under F. B. Weeks (No. 170) were continued in 1905. These grants were clearly in accord with the wishes of Andrew Carnegie, who was supported by the Executive Committee, to discover and develop the exceptional man.

Advisory Committee report

The Advisory Committee on Geophysics for CIW consisted of three geologists, T. C. Chamberlin (University of Chicago), C. R. Van Hise (University of Wisconsin), C. D. Walcott (USGS), and three physicists, R. S. Woodward (Columbia University), C. Barus (Brown University), and A. A. Michelson (University of Chicago) (Figure 1.1).[14] The Advisory Committee chaired by Woodward submitted a list on September 23, 1902 (also published in the first *Year Book*), of sixteen specific problems involving the broader scope of geophysics – as viewed today, encompassing the atmosphere, oceans, and lithosphere – and explicitly including geochemistry. Suggestions for specific research projects and support had been obtained by letter from Lord Kelvin, E. Suess, F. Becke, O. Kohlrausch, J. H. van't Hoff, G. H. Darwin, and W. Nernst. Their list of specific problems (abbreviated here) included:

- heat transfer in the atmosphere
- determination of gases in magmas, rocks, and meteorites
- function of the ocean as a reservoir of atmospheric constituents
- physical chemistry of natural solutions as related to ore deposits
- alteration and recrystallization of minerals under varying conditions
- heat of formation of natural compounds
- deformation of rocks
- effect of pressure on the melting of minerals, including volatiles
- thermal conductivity of rocks
- elastic constants of rocks under varying conditions
- sources of internal heat on Earth
- relationship of heat distribution to deformation and volcanism
- tidal deformation

Figure 1.1   Members of the Joint Advisory Committee on Geophysics. (Top) Geologists: Charles Doolittle Walcott, Charles Richard Van Hise, and Thomas Chrowder Chamberlin. (Bottom) Physicists: Robert Simpson Woodward (Chairman), Carl Barus, and Albert Abraham Michelson. From E. L. Yochelson and H. S. Yoder, Jr., "Founding the Geophysical Laboratory, 1901–1905," *Geol. Soc. Am. Bull.* 106 (1994), 341. By permission of the National Academy of Sciences and the US Geological Survey.

- Moon–Earth tidal relationship
- density and mass distribution in the Earth
- gravity in oceans and continents.

Because "the trustees were not prepared to act" (*CIW Year Book*, 1904, p. xxxv), further study of the subject of geophysical research, especially in Europe, was assigned to Van Hise. On the basis of those discussions he laid out a four-part program considered by geologists as "most pressing" that included the relation of liquid and solid rocks, minerals and rocks from aqueous solutions, deformation of rocks, and physical constants of rocks.[15] In addition, the opportunity to systematize the seismological investigations of the world was recognized. These areas were backed up with a detailed and definitive outline of experiments for the investigation of igneous and metamorphic rocks prepared by eight petrologists in the interest of promoting the "Science of Petrology."

The Committee of Eight

The report submitted on October 10, 1903, by a Committee of Eight provided even greater detail for the initial program of research for the proposed geophysical laboratory. The Committee consisted of Whitman Cross (USGS), Joseph P. Iddings (University of Chicago), Louis V. Pirsson (Yale), and Henry S. Washington (Private Laboratory), the group now famous for the *CIPW* system[16] of rock classification, published in 1902 (Figure 1.2a). Other members were Frank D. Adams (McGill University), James F. Kemp (Columbia), Alfred C. Lane (Michigan State Geological Survey), and John E. Wolff (Harvard) (Figure 1.2b).[17] It was indeed a distinguished group: four of these petrologists became members of the National Academy of Sciences, and one became a foreign associate. Their suggestions for geophysical research, outlined in some detail, are tabulated in brief in Table 1.1.

There appears to be no written record of who organized the committee or how the group was assembled. (It may be presumed that their suggestions resulted from discussions held in Washington when at least six of the Committee of Eight met at the Washington Meeting of the Geological Society of America between December 30, 1902 and January 2, 1903. Neither Pirsson nor Wolff was registered, but Wolff is listed as having given a paper on 2 January.) The emphasis on physics espoused by Becker and the CIW Advisory Committee on Geophysics thereby evolved toward physical chemistry as promoted by Van Hise and the Committee of Eight.

Generating support for a laboratory

Walcott was the principal advocate for an independent research laboratory with the strong backing of the Advisory Committee and the Committee of Eight. Additional support was gained through letters from outstanding scientists abroad and a resolution was engineered by S. F. Emmons at the International Congress of Geologists in Vienna in 1903. Extended discourses on critical problems and geological issues were arranged by Walcott and were given by Van Hise and Becker at the International Congress of Arts and Science of 1904, held in St. Louis. Personal meetings with Carnegie, however, did not appear to advance the concept of a research laboratory.

Whitman Cross, 1854-1949   Joseph Iddings, 1857-1920

Louis V. Pirsson, 1860-1919   Henry S. Washington, 1867-1934

Figure 1.2a   Members of the Committee of Eight who also participated in the formation of the *CIPW* system of rock classification. *Source*: H. S. Yoder, Jr., "Development and promotion of the initial scientific program for the Geophysical Laboratory." In G. A. Good (ed.), *The Earth, the Heavens and the Carnegie Institution of Washington* (American Geophysical Union, 1994), p. 24. By permission of the American Geophysical Union.

Alfred Church Lane, 1863-1948   James Furman Kemp, 1859-1926

Frank Dawson Adams, 1859-1942   John Eliot Wolff, 1857-1940

Figure 1.2b The remaining four members of the Committee of Eight. By permission of the American Geophysical Union. *Source*: As Figure 1.2a.

Late in 1904, Woodward, Chairman of the Advisory Committee on Geophysics, was elected president of CIW, and the next step seemed inevitable. At the meeting of the Trustees on December 12, 1905, Woodward and Walcott persuaded them to establish the Geophysical Laboratory.[18] Carnegie was not pleased and quickly wrote to express his opinion (Figure 1.3). He was clearly opposed both to extracting "exceptional men" from their own environment

Table 1.1   *Geophysical investigations suggested by CIW Committee of Eight, 1903.*

Igneous rocks	Metamorphic rocks
1. Physical properties	1. Physical properties
2. Mutual solution of minerals	2. Thermal properties
3. Diffusion in liquids and solids	3. Rock-water interaction
4. Crystallization from liquids	4. Hydrothermal mineral solubility
A. Liquidus of simple systems	5. Chemical reactions and crystallization
Rates of crystallization	6. Hydration and dehydration
B. Gas solubility in magmas	7. Crystal growth
Hydrous mineral stability	8. Solution of stressed crystals
Crystal size, habit, texture	9. Rock deformation
5. Chemical analysis	10. Development of foliation
6. Thermal properties of minerals	11. Effects of stress on composition
	12. Origin of graphite in metamorphic rocks

Source: *CIW Year Book* 2 (1903), 195–201.

and especially to erecting buildings. Nevertheless, the decision was in the hands of the Board of Trustees, not the donor of the endowment.

Obviously, Woodward and Walcott were so sure of the outcome that major decisions had been made well in advance of the December 1905 vote. What skillful politicians they were! A letter dated October 25, 1905 shows the letterhead already prepared (Figure 1.4) – well in advance of the vote – in the same script used today. It is also apparent that the appointment of A. L. Day, then assistant to Becker at the Survey, as the first director had been made, and yet the old address of the USGS was retained. This simple letter has quite a story behind it,[19] especially in regard to the apparent bypassing of Becker as director, presumably because of his rigid stance on his proposed budget considered excessive by Walcott. Becker has been referred to erroneously as the director by Williams,[20] and in recent editions of the *Encyclopaedia Britannica* (e.g. 1977, vol. 1, p. 981) as the first director of the Geophysical Laboratory. Becker, age 58, was a geologist with a strong physics background, whereas Day, age 36, was a physicist with essentially no geological background. The official appointment of Day as director was made several months into 1906; however, Day had been receiving from CIW a small "honorarium" from April 1, 1904 and a "salary" from April 1905, so the appointment was not unexpected. The work of Day, Allen, and Iddings on the plagioclase feldspars was published by CIW in 1905[21] with the consent of Walcott, CIW secretary and director of the USGS, in recognition of the support from CIW. It was considered to be the first publication of the Geophysical Laboratory, with an extensive introduction written by G. F. Becker!

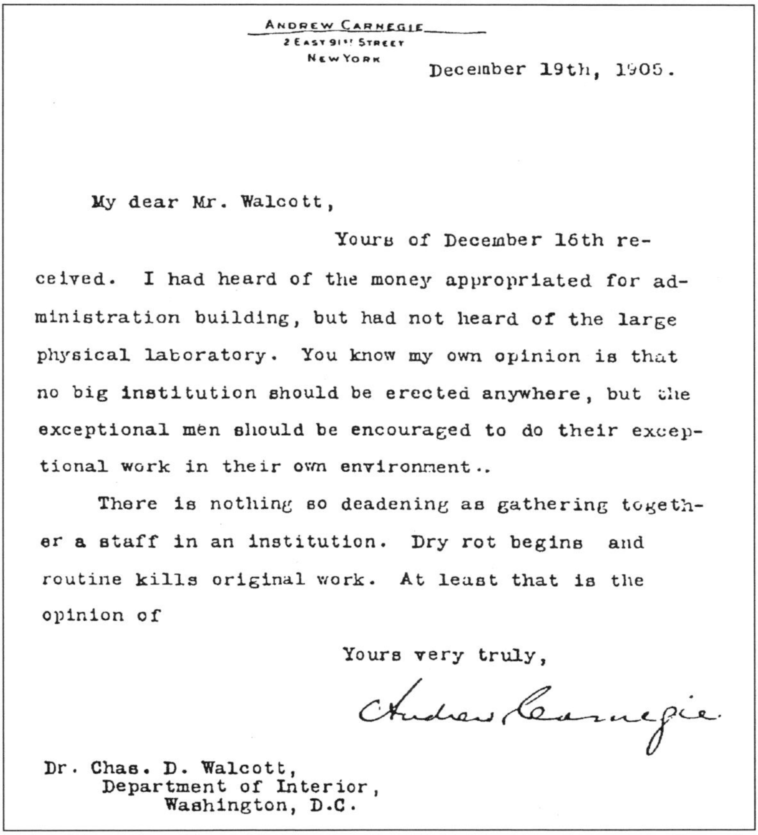

ANDREW CARNEGIE
2 EAST 91st STREET
NEW YORK
December 19th, 1905.

My dear Mr. Walcott,

    Yours of December 16th re-
ceived. I had heard of the money appropriated for administration building, but had not heard of the large
physical laboratory. You know my own opinion is that
no big institution should be erected anywhere, but the
exceptional men should be encouraged to do their exceptional work in their own environment..

   There is nothing so deadening as gathering together a staff in an institution. Dry rot begins and
routine kills original work. At least that is the
opinion of

      Yours very truly,

      Andrew Carnegie

Dr. Chas. D. Walcott,
  Department of Interior,
   Washington, D.C.

Figure 1.3 Letter from Andrew Carnegie to Charles D. Walcott dated December 19, 1905.

Philosophy of approach

Day attributes the philosophy of approach to the problems outlined for the new laboratory to Becker, who said, "We must patiently begin with the simplest problems that can be devised and, aided with the most perfect appliances known, study them exhaustively before proceeding to more difficult and complex cases."[22] In brief, the concept of examining the effects of a single variable while holding all others constant was the cornerstone of the approach. The decision was made to use only the purest chemicals, calibrated temperature and pressure scales, and the component-by-component method of studying phase equilibria. On this basis, Becker directed the initial work at the USGS on the pioneering study of the plagioclase feldspars and some simple eutectic systems. Here, then, was the introduction of the quantitative methods of physics and chemistry in the solution

Geophysical Laboratory
CARNEGIE INSTITUTION OF WASHINGTON

ADDRESS
1330 F STREET
WASHINGTON, D.C.

GEORGE F. BECKER
ARTHUR L. DAY

October 25, 1905.

Dr. Arthur L. Day,

Director Geophysical Laboratory.

Sir:

As noted in your letter of October 16th, the result of a verbal consultation has been that the research on elasticity under my charge has gradually incurred an indebtedness to the machine-shop under your charge, amounting approximately to $500.00. I enclose herewith my check to you for that amount in satisfaction of this debt.

I entirely concur in your suggestion that hereafter an account be kept of work done for me, and that I be charged with the actual time spent upon it and with the actual cost of materials used.

Very respectfully,

Geo. F. Becker

Geophysicist.

Figure 1.4 Letter from George F. Becker to Arthur L. Day dated October 25, 1905.

of geological problems. That philosophy has been maintained throughout the life of the Geophysical Laboratory. Day's experiences during 1897–1900 at the Physikalische-Technische Reichsanstalt, Berlin (founded 1887), no doubt contributed measurably to the successful implementation of this approach.[23]

Laboratory building funded

Although Carnegie had expressed his disapproval of funding permanent buildings, the trustees voted $150,000 to establish the Laboratory, including $100,000 to purchase land and construct a building and $50,000 for equipment. The location of the US Bureau of Standards in northwest Washington greatly influenced Day and CIW president Woodward as to the most suitable area. Woodward selected a 5-acre tract, on an isolated hill in the subdivision known as Azadia, on Randolph Street (later called Upton Street), owned by the estate of Pierce Shoemaker.[24] The adjoining property had just been purchased by the Sisters of the Holy Cross for a school. The land was bought on

March 17, 1906 for $17,500.[25] Of special concern to Day was the freedom of the site from both mechanical and magnetic disturbances. The Upton property was 1700 feet from the nearest streetcar line and the rock foundation did not have a strong magnetic field. In Berlin, Day had observed the effects of the streetcars passing in front of the Reichsanstalt on the very sensitive galvanometers used for measuring temperature.

Building design

The Advisory Committee on Geophysics asked Becker to prepare a detailed estimate of the costs for the proposed laboratory, but he went further than this and also outlined the general character of the building.[26] Becker said "that the experience of the Reichsanstalt is most valuable" (p. 55). In March 1903, Walcott asked Becker to cooperate with Professor Van Hise in gathering more information for the construction of a laboratory building.[27] Apparently the resulting plans were considered too grandiose, so Walcott asked Day in the fall of 1905 to prepare other plans. Day prepared sketches for a three-story building, with an isolated basement floor for the machine shop, storage, battery rooms, heavy high-pressure equipment, and a power plant. In addition, he specified a main floor with a library and independent laboratory units, a second floor for administrative, optical, and photographic work, and living quarters for assistants and visiting investigators so that experiments requiring 24-hour attention could be monitored. The plans drawn up by the architects Wood, Donn, and Deming, who had special expertise in laboratory design, followed Day's specifications (Figure 1.5). Excavation began on June 1906 and the contract for construction was let to Richardson and Burgess, Inc. of Washington, DC on July 6, 1906. A building permit was issued on August 7, 1906 (Historic Landmark Application Form, p. 6). The Spanish Renaissance Revival style was considered well suited to the "climate and the purse of Washington."[28] The author believes the design (Figure 1.6a) was greatly influenced by the design of the Magnus-Haus (Figure 1.6b), headquarters of the German Physical Society in Berlin,[29] although a local architectural historian disagrees. The German Physical Society inaugurated the concept of collaboration of diverse experimentalists in close proximity.

The new building (Figure 1.7) was occupied on June 7, 1907, well before the predicted date of completion (July 1, 1907). The building design served the needs of the Geophysical Laboratory until 1990, requiring only the addition of a small adjoining lead-lined building for X-ray crystallography and two temporary buildings for special war-time projects (see Chapter 20).

Staff recruitment

The staff was primarily recruited from the USGS. Day, as a USGS member, received grants for research in 1904 and 1905, and several staff were actually

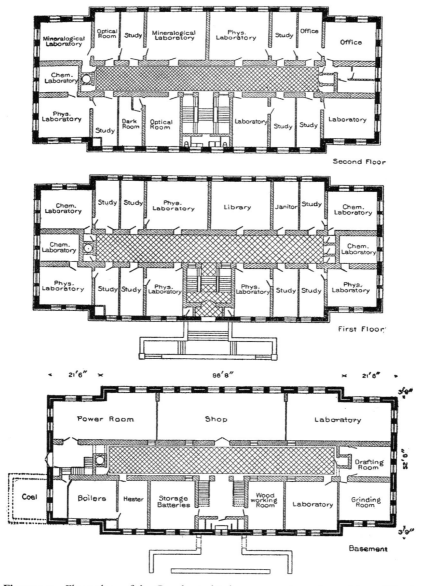

Figure 1.5 Floor plans of the Geophysical Laboratory as drawn by architects Wood, Donn, and Deming. From *CIW Year Book* 6 (1908), plate 7, p. 86 facing.

Figure 1.6a Geophysical Laboratory in 1907 (Archives, GL).

Figure 1.6b Magnus-Haus, 1958. By permission of the Deutsche Physikalische Gesellschaft.

Figure 1.7 Photograph of Geophysical Laboratory under construction in November 1906. From E. S. Shepherd's photo album bearing his title "A house of dreams untold" (Archives, GL).

listed as members of the Geophysical Laboratory even before it was officially formed on December 12, 1905! After Day was named director on January 1, 1907, the staff consisted of the following:

Staff member	Position	Date of GL employment
E. T. Allen	Chemist	1 Jan 07
W. Beck	Mechanician	1 Jan 07
B. D. Chamberlin	Master Mechanic	6 June 04
J. K. Clement	Physicist	15 Nov 04
A. L. Day	Director	1 Jan 07
C. W. H. Ellis	Technical Assistant	6 June 04
G. R. Hoffman[a]		
E. S. Shepherd	Physical Chemist	1 Dec 04
A. F. Susan	Stenographer	1 Jan 07
W. P. White	Physicist	1 Aug 04
F. E. Wright	Petrologist	1 Jan 06

Note: [a] In the pay ledger, G. R. Hoffman is listed as a temporary employee only for January 1906.

A photograph (Figure 1.8) was taken of the group by Shepherd, presumably in January 1906, when they were still working on the fourth floor of the USGS building at 1330 F Street, NW, Washington, DC.

Figure 1.8 Photograph of initial staff members of the Geophysical Laboratory while located at the US Geological Survey, 1330 F Street, NW, Washington, DC. Front row, left to right: G. R. Hoffman, E. T. Allen, A. L. Day (Director), and W. P. White. Back row, left to right: W. Beck, C. W. H. Ellis, F. E. Wright, B. D. Chamberlin, J. K. Clement, and A. F. Susan. Photograph from Shepherd's album (Archives, GL) and presumably taken by him.

Scientific program outlined

The following broad program of research outlined in 1902 was reviewed and rededicated by Day in 1927.

> The crust of the lithosphere has thus far been the chief field of geology in the narrower sense, since it contains the rock record of the Earth's past; and geological studies have been directed chiefly to reading and mapping this record, but the record needs to be interpreted on broader and deeper lines based on a profounder knowledge of physical laws. To this end, the data of geology need to be correlated and unified under these laws on an experimental basis . . .
>
> Some of the salient problems of the outer lithosphere are the origin and mainten-ance of the continental platforms . . . and a whole group of intricate questions of a chemical and chemico-physical nature, including the flow of rocks, the destruction and genesis of minerals, the functions of included water and gases, the internal transfer of material, the origin of ore deposits, the evolution and absorption of heat, and other phenomena that involved the effects of temperature, pressure, tension and resultant distortion upon chemical changes and mineralogical aggregations.

These questions of the Earth's outer part are inseparably bound up with those of the interior, and here the problems involve the most extreme and the least known conditions and make their strongest demand for experimental light. The themes here are the kinds and distribution of the lithic and metallic materials in the deep interior; the states of matter, the distribution of mass and of density, and the consequent distribution of pressure; the origin and distribution of heat . . . the secular redistribution of heat within the Earth and its loss from the surface; the possible relations of redistribution of internal heat to vulcanism, and to deformation and similar profound problems.

A series of specific laboratory questions arise from these, e.g., the effect of pressure on the melting point of rocks carried to as high temperatures and pressures and through as wide range of materials as possible to develop the laws of constancy or of variation; the effect of temperature and pressure on thermal conductivity as indicated above, and on elasticity, especially as involved in the transmission of seismic tremors.

Over the years, more than 3100 papers have been issued from the Geophysical Laboratory, but this represents only one measure of the contribution of the staff. The ebb and flow of the focus of the work has indeed been great. Figure 1.9 gives a crude picture of the change in effort of the regular staff throughout the years.[30] The effort assigned to various fields is somewhat arbitrary in view of the overlap and integration of the fields. The designated fields, however, serve as focal points on which to summarize the following highlights of the work in subsequent chapters. It is evident that the Geophysical Laboratory has responded dynamically to the needs of the science, developing the most rewarding directions as they evolved. It is also evident that it is the individual staff members who have made the concept of interdisciplinary research so successful. In accord with the original wishes of Mr. Andrew Carnegie, the support of exceptional individuals working cooperatively has resulted in a record of discovery and invention that is extraordinary.

Consolidation and/or collocation

On March 22, 1983, the then president of CIW, James D. Ebert, invited the directors of the Department of Terrestrial Magnetism (DTM), G. W. Wetherill, and the Geophysical Laboratory, H. S. Yoder, Jr., to lunch at the Cosmos Club in Washington, DC. The principal issue presented to the directors was the intention of the president to consolidate the programs of the two laboratories on one geographical site. Although the reasons for this move were never explicitly stated, the president, a biologist, had a low opinion of the earth sciences (expressed publicly at the 1983 Annual Meeting of CIW) and did not distinguish the unique research directions of the two departments. He did note, however, that terrestrial magnetism was no longer studied at DTM and geochemistry appeared to be the main focus at the

Year

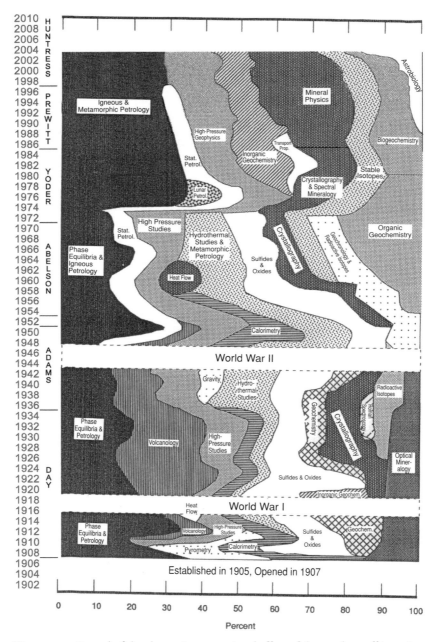

Figure 1.9 Record of the change in proportional effort of the regular staff in various fields of research over time and under different directorships. George W. Morey served as acting director, 1952–1953; Robert B. Sosman served as acting director, 1918–1920. Revised from Yoder, "Scientific highlights," p. 146.

Laboratory, but a new descriptive title for the departments was not offered.[31] The age of the buildings of each department was not at issue, and only minor renovation changes had been requested (e.g. a clean laboratory for the new biochemical studies). Collaboration on studies of shared interests had taken place as needed (see chapters on seismology, geochronology, computation and isotope geochemistry). The move was perceived to be a method to reduce the number of departments and the size of the staff, thereby reducing the need for additional operating funds. By September 1983 the president did request, however, that the Geophysical Laboratory prepare a statement of the staff's perceptions of the scientific directions over the next ten years with specific short, medium and long-range research goals.

Needless to say, the response of the staff in each department was immediate and vigorous. A survey of staff members produced a unanimous and strong feeling that the Geophysical Laboratory should maintain its identity, which would be blurred or lost if merged with DTM. The necessity of leaving the Washington area was a primary family concern, but a move to a research university campus presented many advantages. A university location would provide access to experts in a wide range of basic sciences. There would be a greater probability of interdisciplinary studies enlarged by spontaneous cooperation. Economies might be achieved with shared facilities as well as access to new facilities. Students could supplement the limited number of postdoctoral fellows in expediting the research. The opportunity for teaching might focus the research efforts on more critical problems. Within a year, written indications of interest were received from Cornell University, University of Chicago, The Johns Hopkins University, MIT-Harvard University (joint proposal), Pennsylvania State University, Princeton University, and Stanford University. Verbal expressions of interest had also been made by Georgia Institute of Technology and University of Maryland. In addition to Committees organized to examine the criteria for consolidation and/or collocation, a joint Planning Feasibility Study Committee met on March 5, 1984 to focus on the advantages and disadvantages of such a move. Both the president, J. D. Ebert, and vice-president, M. L. A. MacVicar, of CIW attended, and additional discussions were held individually and jointly throughout 1985. Visits were made to and by the interested universities. On May 3, 1985 the Board of Trustees "Resolved that, because of the desirability of locating the Institution's programs in the Geophysical Laboratory and the Department of Terrestrial Magnetism onto a common site, the Institution proceed with plans for new or remodeled buildings for the departments at a single site . . ."[32] In spite of the opposition of the directors of the two departments and their staff members, the move to a common site was "with the expectation of consolidation of the two departments" (p. 166). The "forthcoming marriage" was alleged to be an "evolutionary response to today's opportunities" (p. 5). In short, a reduction in staff numbers was expected, as was an increase in pressure on staff for obtaining additional funding for its projects. Because of the anticipated changes in leadership of the Carnegie Institution (July 1, 1987) and the Geophysical Laboratory (July 1, 1986) from mandated retirements, there was adequate time for discussion of plans with architects. After reviewing alternative proposals for the future consolidation of the Geophysical Laboratory and DTM, on May 9, 1986, the Board of Trustees "voted authorization to commission an architectural schematic design for new construction and renovation of existing buildings at the campus of DTM in northwestern Washington."[33]

New building

The architects of the firm Peirce, Pierce & Kramer of Cambridge, MA, were commissioned as early as April 1983 to assist in planning the consolidation of the three CIW properties in the District of Columbia, which initially included the Departments of Administration, Terrestrial Magnetism, and the Geophysical Laboratory. They were assigned the following tasks in the "Planning Feasibility Study":

1. To identify the present and future space needs of the Office of Administration, the Geophysical Laboratory, and the Department of Terrestrial Magnetism.
2. To report on the existing condition of the buildings and sites presently occupied by these separate departments.
3. To examine the various planning options for meeting both individual departmental needs and those of the Carnegie Institution in the most effective way.
4. For the option (or options) selected, to develop schematic design proposals, construction cost estimates and project schedules as the basis for subsequent action by the Board of Trustees.

By January 1984 the architects were advised by Vice-President MacVicar of CIW to concentrate on consolidation of GL and DTM, with the Administration remaining at its present site. The functional groups were identified, the support functions listed, and special room requirements detailed after the staff completed written questionnaires. The Planning Feasibility Study was submitted on February 1, 1985, which included a site model of the preliminary design proposal at the DTM site (Figure 1.10). Only the Standard Magnetic Observatory and a maintenance shed were sacrificed for the new building. Special care was given to minimize the "institutional" impact of the development on the residential neighbors of the DTM site.

Response to the preliminary designs was extensive because of the many special requirements of the wide variety of research programs. The acting president of CIW, Edward E. David, Jr., found it necessary to hire an experienced

Figure 1.10 Model of the new research buildings as presented by the architects February 1, 1985. Clockwise from lower left: Main building of DTM, Experiment building, Accelerator building, Cyclotron building, maintenance shed, new research laboratory. From Peirce, Pierce and Kramer, Inc., *CIW Planning Feasibility Study*, 1985, p. 14.

Figure 1.11 Groundbreaking ceremony January 31, 1989 for the new research buildings on the DTM Campus. Left to right: Richard Heckert (Chairman of the Board of Trustees), Maxine Singer (President CIW), Jack Clogett (Senior Vice-President, Donohoe Construction Co.) and Robert Kramer (architect, Peirce, Pierce, and Kramer, Inc.). From *CIW Newsletter*, March 1989, p. 3.

project manager, Murray Stewart, from Exxon on November 16, 1987 to facilitate the changes. The Joint Planning Committee insisted that the changes requested were based on the experiences of the professional experimenters. Many of the critical detailed changes were brought to fruition mainly through the special efforts of one staff member in particular, T. Neil Irvine.

Ground was broken on January 31, 1989 (Figure 1.11) for the new research building and the ceremony was favored by remarks from the president of the National Academy of Sciences, Frank Press:[34]

> In the early days, we were in awe of these laboratories. In fact, our mnemonic code for the Geophysical Laboratory was the "Gee Whiz" Laboratory. But perhaps the best compliment I can pay is the fact that, as a department chairman at MIT, we made offers to literally half a dozen of the staff members, and none of them came. They liked it here too much.
>
> What we witness today – the beginning of construction of a new laboratory – demonstrates that the Carnegie Institution continues to fulfill Andrew Carnegie's vision of the essentiality of new knowledge and its contribution to the American enterprise and, for that matter, to that of all nations.

Figure 1.12 Photograph of the new research building on the DTM Campus, 1990 (Archives, CIW).

The new building at 5251 Broad Branch Road, N. W. (Figure 1.12) was completed in May 1990 by the Donahoe Construction Co. The Geophysical Laboratory occupied the two eastern wings and DTM the western wing. The administrations of the two departments were housed in the renovated old main building of DTM and the libraries consolidated on the second floor and attic. To insure accessibility to the administration staff and the laboratory, a tunnel (Figure 1.13) was constructed between the new building and the old building. Provision was also made for lecture and conference rooms, as well as offices for visitors and postdoctoral fellows. In response to concerns about electrical breakdowns, independent circuits were run to each laboratory (Figure 1.14).

The old GL building on Upton Street was declared a Historic Landmark of the National Capital on December 29, 1994 through the efforts of the concerned neighbors and the support of the Upton Street Preservation League president, Emma C. Jordan. The historical landmark was declared on the grounds that it was (1) a laboratory of national importance; (2) its contributions were known world-wide; (3) the structure was an important product of a noted local architectural firm; (4) the site is associated with two men (Joshua Peirce and Pierce Shoemaker), prominent in local history; and (5) the site has been retained as an open area for specific scientific purposes. This declaration greatly restricted changes to the building and made it difficult to find a qualified buyer. Interested buyers included the Finnish Embassy, American Geophysical Union, Cornell University, Washington Episcopal School, and

Figure 1.13   Tunnel constructed below ground level for all-weather access between renovated administration building (upper right) and new research building (lower left). Photo by G. Bors.

Figure 1.14   Electrical conduits in first floor slab of the new building to insure independent electrical supply and avoid cascading of breakdowns. Photo by G. Bors.

the Levine School of Music. The property was eventually sold for $2,300,000 on December 18, 1995 to the Levine School of Music.[35]

Interactions from collocation

In the dozen years that have passed since collocation many new friendships have been formed and a broader appreciation of the work of each department has evolved as a result of ready access to seminars held by each department. Of particular significance is the cooperation on problems involving isotope geochemistry. In addition, the Geophysical Laboratory and DTM both became part of the NASA-sponsored Astrobiology Institute to investigate not only the origin of life on Earth but also the potential for life on other planets. Study of meteorites continues as a cooperative venture of the two departments. The extensive requirements for advanced computation have also deeply involved both groups. Needless to say, critical review of the entire program from external associate investigators outside one's field has been of value to the staff. Some of the most productive links between the departments has been through interests of the postdoctoral fellows and visiting investigators.

2

IGNEOUS PETROLOGY

The charter for the Geophysical Laboratory, as recorded in the report of the Advisory Committee on Geophysics, clearly stated the need to unify geological field observations under physical and chemical laws on an experimental basis. The Committee believed geologists wanted to know:

> the melting points of rocks, the temperatures at which rocks crystallize from magma, the relative specific gravities of melted and crystallized rocks, the effects of slow cooling upon the crystallization of rocks with and without pressure, the solution of one kind of rock in another, and in short, all the phenomena which concern the transformation of magma to crystallized rock and of crystallized rock to magma.

That statement appeared in *CIW Year Book* 2 for 1903, which included a detailed plan of investigation (pp. 195–201). It has served as the principal guideline for the core program of the Geophysical Laboratory for almost a century.

Temperature scale

The experimental approach to those goals was immediately beset with problems of the most fundamental nature. There was no generally acceptable temperature scale above 200 °C and standard calibration points had not been established even though several boiling and melting points were commonly used up to about 1100 °C.[1] There was, however, considerable knowledge about the composition of rocks, and the ten most important oxides had been identified by chemical analyses (Table 2.1). The main advantage lay with the incredible intuition and perception of the geological advisors who had acquired a remarkable qualitative sense about how rocks were formed.

The most abundant mineral in the crust of the Earth is plagioclase, and the albite (Ab) – anorthite (An) system had been selected for study at the US Geological Survey in the formative years of the Geophysical Laboratory. That first step in a much broader scheme of investigation of the common rock-forming minerals was undertaken by Day, Allen and Iddings (University of Chicago) with the financial support of CIW. Their results on the Ab–An system were published in 1905 as paper No. 1 of the new Geophysical Laboratory.[2] The liquidus was determined by the heating curve method

Table 2.1 *The chemical and geological composition of the crust*

A. Chemical composition of crust (average wt %)

SiO_2	59.3	Fe_2O_3	2.5
Al_2O_3	15.9	K_2O	2.4
CaO	7.2	TiO_2	0.9
FeO	4.5	P_2O_5	0.2
MgO	4.0	MnO	0.1
Na_2O	3.0		100.0

B. Abundance of rock types in crust (average volume %)

Basalts	42.5
Gneisses	21.4
Granodiorites	11.2
Granites	10.4
Crystalline schists	5.1
Clays and shales	4.2
Carbonates	2.0
Sands	1.7
Marbles	0.9
Syenites	0.4
Dunites and peridotites	0.2
	100.0

whereby the melting point is marked by a single break in the heat curve from An_{100} to $An_{26.1}$ (Figure 2.1). The remainder of the now classical solid solution loop was deduced as Roozeboom's Type I. The temperature calibration was based on the Reichsanstalt scale for the melting of Cd, Zn, Hg, and Cu. Platinum–rhodium thermocouples were employed with a Pt-wound resistance furnace.

Quenching method

Because of the great difficulty in determining the exact temperature of complete melting, Shepherd, Rankin, and Wright devised a new method in 1909 for the $CaO\text{-}SiO_2$, $MgO\text{-}SiO_2$, and $Al_2O_3\text{-}SiO_2$ systems in which the liquid was quenched from a known temperature to a glass and examined optically for crystals.[3] A small packet of a mixture of the end members was held at constant temperature and then let fall into a dish of mercury below the vertically oriented furnace. The abrupt chilling fixes the kind and composition of the phases grown at the high temperature. In this way, the limits for the stability fields of the phases could be worked out by making a series of runs at various temperatures for each mixture. This new quenching technique was applied in a re-examination of the plagioclase system in 1913 by N. L. Bowen,[4] who

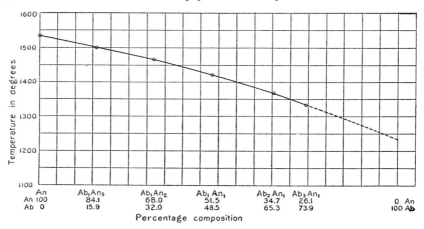

Figure 2.1   Albite (Ab)–anorthite (An) liquidus determined by the cooling curve method. *Source*: From A. L. Day, E. T. Allen, and J. P. Iddings, *The Isomorphism and Thermal Properties of the Feldspars* (Carnegie Institution of Washington Publication 31, 1905), p. 60, figure 14.

proved that the solid solution loop was indeed as deduced (Figure 2.2).[5] He also showed that the depression and rise of the melting temperatures of the endmembers An and Ab, respectively, were in close agreement with Raoult's Law of vapor pressure. Thus, plagioclase may be considered an ideal solution; however, the conditions under which it deviates from ideality remains a principal focus today. The superiority of the quenching method over the heating method for silicates was demonstrated by Morey in 1923.[6]

Geophysical Laboratory temperature scale

The new Geophysical Laboratory temperature scale, calibrated with lithium metasilicate (Li_2SiO_3), diopside ($CaMgSi_2O_6$), and anorthite ($CaAl_2Si_2O_8$), was applied in the experiments described above. The very first practical problem to be faced by the staff of the Geophysical Laboratory was the calibration of a temperature scale above about 1100 °C, there being no internationally accepted scale in 1905. Because of Day's experience at the Physikalisch-Technische Reichsanstalt in Berlin, the first apparatus installed was the nitrogen-gas thermometer.[7] With a new design for improved accuracy of measurement, Day and Sosman[8] provided accurate data for a scale from 300 °C to 630 °C with a direct determination of the boiling point of sulfur. The scale was then extended by them to the palladium melting point of 1549.2 °C. The fixed-points were corrected from the constant-volume nitrogen scale to an absolute thermodynamic scale and expressed by Adams[9] as an e.m.f. for copper-constantan and platinum-rhodium thermocouples from 0 °C to 1755 °C. The principal fixed points for gold, copper, diopside,

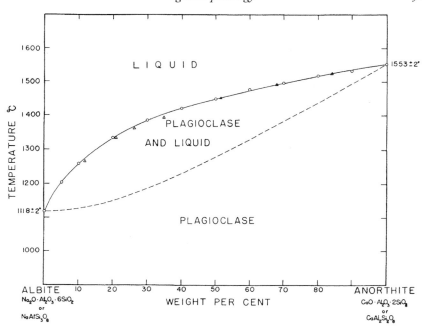

Figure 2.2 Albite–anorthite system determined by the quenching method by Bowen (triangular points) in 1913 and by J. F. Schairer (open circles) about 1948. From H. S. Yoder, Jr., "Experimental mineralogy: achievements and prospects," *Bull. Mineral.* 103 (1980), p. 10, Figure 10. By permission of the *European Journal of Mineralogy*.

and palladium became known as the Geophysical Laboratory scale,[10] and it is still used today. The melting point of platinum, 1755 °C, was based on the incremental difference of optical pyrometer measurements of the Bureau of Standards between palladium and platinum.

International Temperature Scale

After World War I an international conference met and the International Temperature Scale of 1927 was adopted. Silver was set 0.3 °C higher; gold 0.4 °C higher, palladium 5.5 °C higher, and the major change was that platinum became 1773 °C. In spite of these recommendations, the scientists at the Geophysical Laboratory retained their own scale because of their belief in its thermodynamic foundation and for consistency with their previously published phase diagrams. In the meantime, physicists noted certain discrepancies in the physical constants related to the International Temperature Scale. A new international committee met in 1939, but World War II intervened, and their results were not published until 1948. The changes nearly restored the fixed values below 1550 °C to those of the Geophysical Laboratory

Figure 2.3   Bank of resistors (lower left corner of photograph) for controlling the temperature of an electric furnace in the laboratory of Day and Sosman in 1911. From Day, Sosman, and Allen, *High Temperature Gas Thermometry*, frontispiece.

scale. One significant change related to the extrapolation to the melting point of cristobalite which would require a change from 1713 °C to 1723 °C. Nevertheless, the values for diopside (1391.5 °C), pseudo-wollastonite (1544±2 °C), and cristobalite (1713±5 °C) have all been retained by the Geophysical Laboratory as calibration points not only for consistency but also because they are within the errors of experimental determination of the International Temperature Scale, which is still subject to change. It would be useful to ascertain in the near future the melting point for the important endmember mineral forsterite, which is still known only as 1890±20 °C. The corrections to the e.m.f. of thermocouples used at high pressures, recently studied but subject to debate, also remain an important subject for reinvestigation. In principle, the triple points of mineral phase changes may serve as an adequate P–T scale for interlaboratory comparisons.

Temperature regulation

Maintaining the temperature of a furnace was also a challenging problem for the experimenter. Initially, the electric furnaces were driven by a large number of wet battery cells, and temperature control was achieved by a very large street-car-type set of resistor coils (Figure 2.3) that were manually

Figure 2.4   Early design of the Roberts temperature regulator, above Dr. J. F. Schairer's head on right-hand side of photograph. Six quenching furnaces are displayed. The emf of the thermocouples is read on the potentiometer. (Photo. 1950 by J. Harper Snapp; Archives, GL).

adjusted. Needless to say, 24-hour runs were a strain on the operator who had to make adjustments continually as the ambient temperature changed. In 1919, White and Adams[11] made the heating coil of the electric furnace as one arm of a Wheatstone bridge. By combining this with a galvanometer, it was possible to keep the resistance of the coil constant regardless of variations in the current supply, thereby maintaining the temperature of the furnace constant. Improvements made by Roberts in 1921[12] meant that linear heating rates could be achieved, and the regulator could be adapted to alternating current (1922). He claimed in 1925[13] that with a triode tube relay the regulator could be operated continuously for a month without any attention to 0.1 °C.

A large number of phase diagrams were determined accurately with various modifications (1940) of the Roberts regulator (Figure 2.4). Where less accuracy was required, for example in preparing large volumes of new mixtures, a variac was used to control the voltage. By 1968 solid-state temperature controllers had been designed by C. Hadidiacos[14] and integrated into both one-atmosphere (Figure 2.5) and high-pressure types of apparatus. Full control of all heating segments and log maintenance was achieved by Hadidiacos

Figure 2.5   Solid-state temperature regulators using a variac were employed after 1968 by Dr. J. F. Schairer. Dr. Schairer is holding the 110 volt quenching device for melting the wire that holds the ceramic ring and attached charge in the center of the furnace (Archives, GL).

and D. George using a Commodore Computer with 64K memory,[15] and by 1990 an IBM interface board was designed to control up to four furnaces. That design is now used throughout the Geophysical Laboratory to control temperature, ramping, stepping, or intricate heating processes.[16]

Physico-chemical methodology

By systematically determining the physico-chemical relations in the binary systems of paired oxides initially, it is possible to add additional components, thereby approaching the ten-component natural system of the rocks themselves. Most of the 45 binary oxide systems have been done at ambient pressure, and a large number of the potential 120 ternary systems are determined, but only a dozen of the 210 quaternary systems have been attempted. The purpose in presenting some of these detailed diagrams (Figures 2.6 through 2.13) is to document the method of approach in establishing how various igneous rock types, that is, those crystallized from a liquid (magma), are related to each other. A large amount of labor is required to establish

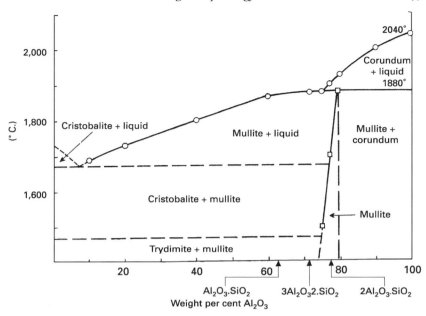

Figure 2.6 The binary system SiO_2-Al_2O_3 at one atmosphere. From J. H. Welch, "A new interpretation of the mullite problem," *Nature* 186 (1960), p. 546, Figure 2. Reprinted with permission of Nature Publishing Group.

these relationships, but the temperatures observed closely approximate those observed when similar natural rocks themselves are melted and crystallized. All of the diagrams were initially determined at one atmosphere, and the principal task underway is to examine each diagram at a series of pressures to establish the changes in minerals and their assemblages with depth in the Earth (see Chapter 3 on Pressure). The logical and systematic investigation of igneous rocks in this way has been one of the great successes in geological research in the last century.

Oxides vs. normative minerals

The most abundant oxides in the Earth's crust are SiO_2 and Al_2O_3. Figure 2.6 gives their binary behavior as a function of temperature.[17] In Figure 2.7, CaO has been added to form a ternary system.[18] Its study was particularly rewarding because it led to an understanding of many ceramics, portland cement, and slag formation in steel making. Another ternary diagram, Na_2O-CaO-SiO_2,[19] has served as the basis for the commercial soda-lime glass industry. Within a period of about fifty years, L. H. Adams listed some 300 systems and subsystems experimentally determined at the Geophysical Laboratory.[20] On

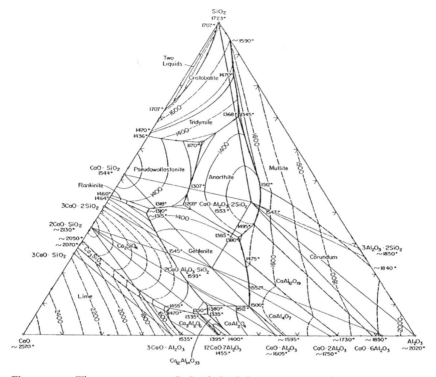

Figure 2.7   The ternary system $CaO-Al_2O_3-SiO_2$ at one atmosphere. From E. F.
Osborn and A. Muan, *Phase Equilibrium Diagrams of Oxide Systems* (American
Ceramic Society, 1960), plate 1. Reprinted with permission of the American Ceramic
Society.

the 75th anniversary in 1981, R. M. Hazen and M. H. Hazen indexed all
the Annual Reports of the Laboratory and the list of investigated systems
totaled just over 400.[21] Whereas the initial focus was on oxide systems, the
interest turned to normative minerals. This change was no doubt influenced
by Bowen's thesis[22] and articles on the evolution of the igneous rocks.

Reaction principle

In the short space of ten years, Bowen had at his disposal the data for systems nepheline-anorthite (Ne-An),[23] albite-anorthite (Ab-An),[24] diopside-
forsterite-quartz (Di-Fo-Qz),[25] anorthite-forsterite-quartz (An-Fo-Qz),[26]
diopside-albite-anorthite (Di-Ab-An),[27] and $CaO-Al_2O_3-SiO_2$.[28] From these
data and a large measure of genius, Bowen produced "The later stages of the
evolution of the igneous rocks." In 1922, he published the article "The reaction principle in petrogenesis,"[29] which Pentti Eskola of Finland later called

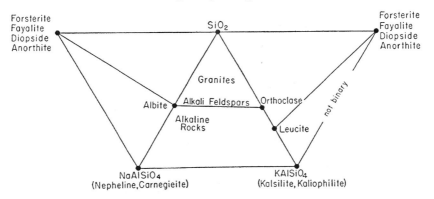

Figure 2.8 Petrogeny's residua system NaAlSiO₄-KAlSiO₄-SiO₂ and related early- and late-forming minerals in a crystallizing magma. *Source*: From *CIW Year Book* 54 (1955), 142, figure 15.

"the most important contribution to petrology of the present century." With only the additional information in CaO-MgO-Al₂O₃,[30] MgO-Al₂O₃-SiO₂,[31] CaO-MgO-SiO₂,[32] akermanite-gehlenite (Ak-Geh),[33] and the immiscibility studies of Greig[34] in FeO-Fe₂O₃-Al₂O₃-SiO₂, Bowen assembled all his previous petrological discussions in a set of Princeton lectures published in 1928 as *The Evolution of the Igneous Rocks*.[35] Although Bowen expressed his great prejudice for Becker's well-established theory of crystal fractionation[36] as the guiding principle in accounting for the diversity of rocks, he also provided the theory for testing alternative views. No other book has had a greater influence on the course of petrology. On the fiftieth anniversary of Bowen's book, a review of the same questions raised by Bowen was published,[37] and it was evident that he had indeed discussed the critical issues still relevant today.

Flow sheet

After the determination of nepheline-kalsilite-quartz system (Ne-Ks-Qz) (Figure 2.8) by Schairer and Bowen in 1935,[38] called "petrogeny's residual system" by Bowen (1937),[39] the course of phase equilibria research was set for years to come. Systematically, Schairer and colleagues added the endmembers of each of the phases formed early in magma (e.g. forsterite, anorthite, diopside, and enstatite) to the relevant joins in the residua system. As the data accumulated, the more difficult it became to relate the direction of falling temperatures, the univariant curves, and the invariant points. This dilemma was resolved by Schairer who devised a "flow sheet" for the system CaO-FeO-Al₂O₃-SiO₂.[40] To illustrate the principle, the data in diopside-albite-enstatite-quartz (Di-Ab-En-Qz) (Figure 2.9)[41] are assembled schematically in a tetrahedron (Figure 2.10)[42] and are expressed in a flow sheet

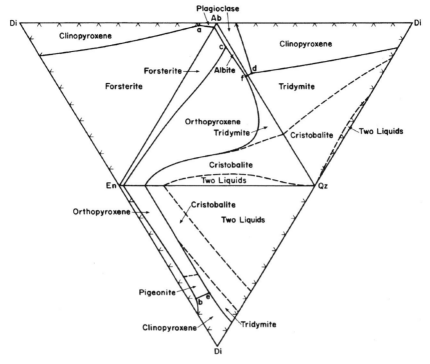

Figure 2.9   The diopside ($CaMgSi_2O_6$) – albite ($NaAlSi_3O_8$) – enstatite ($MgSiO_3$) – quartz (SiO_2) system laid out to show the liquidus relations of the boundary systems at one atmosphere. From H. S. Yoder, Jr., *Generation of Basaltic Magma* (National Academy of Sciences, Washington, DC, 1976), p. 126. Reprinted with permission from the National Academy of Sciences, National Academies Press.

(Figure 2.11).[43] This approach can be applied to more complex systems such as larnite-nepheline-forsterite-quartz (La-Ne-Fo-Qz), the expanded basalt system, containing five oxides.[44] The flow sheet is shown in Figure 2.12, and the rock nomenclature diagram corresponding to the mineral assemblages in Figure 2.13. In this way the interrelationships of the various rock types can be ascertained. It is not evident how the ten-component natural system can be displayed, but matrices have been applied to represent the data.[45] The parent–daughter relationships of the various magmas have also been tested using the natural rocks themselves.[46]

Iron problem

Natural rocks usually contain iron, which occurs in two oxidation states, FeO and Fe_2O_3. In 1915 it was observed by Sosman and Hostetter,[47] in their study of synthetic rock systems, that iron oxides were reduced in platinum

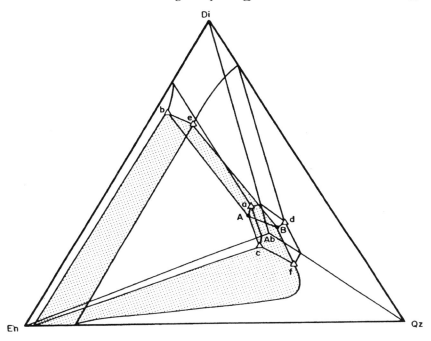

Figure 2.10 Schematic assembly of Di-Ab-En-Qz illustrating probable location of the "quaternary" invariant points A and B. The enstatite volume is stippled as an aid to visualize the three-dimensional relations. From Yoder, *Generation of Basaltic Magma*, p. 127. Reprinted with permission from the National Academy of Sciences, National Academies Press.

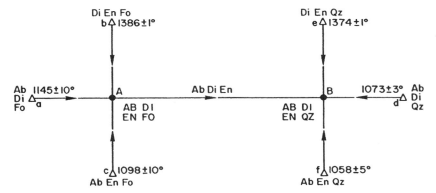

Figure 2.11 Flow sheet for the system Di-Ab-En-Qz displayed in Figures 2.9 and 2.10. Lower-case letters represent "invariant" points in Figure 2.9; upper-case letters, those in Figure 2.10. Arrows indicate direction of falling temperature. The temperature of A is $< 1098° \pm 5$ °C, and the temperature of B is $< 1058° \pm 5$ °C. From Yoder, *Generation of Basaltic Magma*, p. 128. Reprinted with permission from the National Academy of Sciences, National Academies Press.

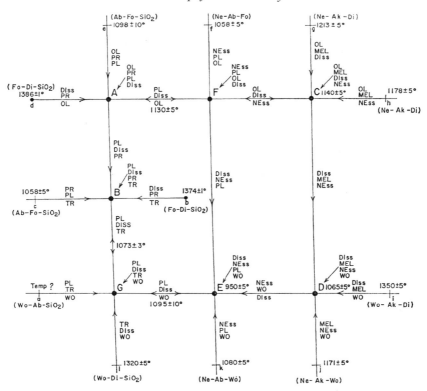

Figure 2.12 Flow sheet determined experimentally for the petrologically important portions of the expanded basalt tetrahedron larnite-nepheline-forsterite-quartz. After J. F. Schairer and H. S. Yoder, Jr., "Crystal and liquid trends in simplified alkali basalts," *CIW Year Book* 63 (1964), p. 72, figure 8.

crucibles. It became evident that it would be necessary to control the P_{O_2} to fix the oxidation state. Use of a vacuum furnace or a reducing atmosphere was not reliable, even when the products were analyzed. Subsequently, in 1932, Bowen and Schairer[48] resorted to holding the mixtures in a pure electrolytic iron crucible in a "neutral" stream of nitrogen. The analysis of the products yielded an equilibria diagram for $FeO-Fe_2O_3-SiO_2$. With the same technique, Schairer[49] determined the critical relations in $CaO-FeO-Al_2O_3-SiO_2$ ten years later. He outlined the entire flow sheet exhibiting eleven quaternary invariant points, two of which are of special petrologic interest because of their potential as parental magmas for the generation of basalt. The mutual melting relations were determined for pyroxenes, pyroxenoids, olivines, and melilites, each mineral group involving extensive solid solutions. No doubt the application of the results were also of exceptional interest to the metallurgists and ceramicists.

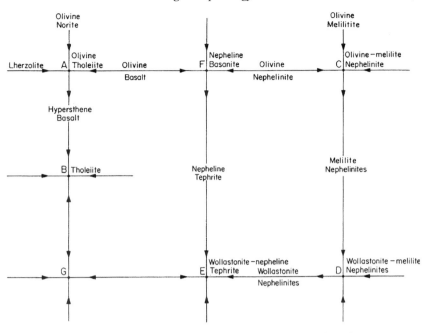

Figure 2.13 Rock nomenclature diagram corresponding to the flow sheet in Figure 2.12. Names are for extrusive rock types. *Source*: As Figure 2.12, p. 73, figure 9.

For systems requiring only Fe_2O_3, Huckenholz and Yoder mixed PtO_2 in the starting materials held in Pt tubes to ensure an excess of oxygen for the study of andradite and ferridiopside at high pressures.[50] Another method for maintaining the partial pressure of oxygen, developed elsewhere,[51] was the use of mixtures of gases, such as CO_2-H_2. Eventually it was possible to define the Mg-Fe fractionation trends for the major rock-forming phases as a function of $f(O_2)$ and T. Other methods for regulating the oxidation state of iron in synthetic preparations are presented in Chapter 4 on volatile components.

A major tool for petrologists was provided by Lindsley,[52] who calibrated the coexisting pairs of Fe-Ti oxides for use as thermometers and oxygen barometers. Because the oxidation state of iron has a profound influence on the differentiation trend of a magma, the quantitative measure of the partial pressure of oxygen with minerals of widespread occurrence has been of exceptional value.

Microscopy

Identification of the fine-grained products, some as small as 10 μm, in synthetic silicate runs imposed a new and difficult problem. New methods

Table 2.2 *Minerals named after Geophysical Laboratory investigators*

Mineral name	Formula	Staff	Position	Dates (m/d/y)
Abelsonite	$C_{13}H_{32}N_4Ni$	Philip H. Abelson	Biogeochemist/Dir.	9/1/53–7/1/71
Adamsite – (Y)	$NaY(CO_3)_2 \cdot 6H_2O$	F. D. Adams	CIW Grants	1903–1911
Bartonite	$K_3Fe_{10}S_{14}$	Paul Barton	Guest Investigator	April–Oct. 1957
Buddingtonite	$NH_4AlSi_3O_8 \cdot 1/2H_2O$	Arthur F. Buddington	Staff Petrologist	8/25/19–9/1/20
Burtite	$CaSn(OH)_6$	Donald M. Burt	Fellow	9/1/70–9/31/72
Chaoite	C	E. C. T. Chao	Guest Investigator	1967–1969
Chayesite	$K(Mg,Fe^{2+})_4 \ Fe^{3+} Si_{12} \ O_{30}$	Felix Chayes	Research Associate	5/16/47–5/15/49
			Petrologist	5/16/49–6/30/73
			Systematic Petrologist	7/1/73–6/30/86 (part-time)
Donnayite – (Y)	$Sr_3NaCaY(CO_3)_6 \cdot 3H_2O$	J. D. H. Donnay	Guest Investigator	1/1955–9/1970
Eskolaite	Cr_2O_3	Pentti Eskola	Research Assistant	3/1/21–12/15/22
Esperite	$(Ca,Pb)ZnSiO_4$	Esper S. Larson, Jr.	Assistant Petrologist	9/1/07–6/1/09
Eugsterite	$Na_4Ca\,(SO_4)_3 \cdot 2H_2O$	Hans Peter Eugster	Guest Investigator	10/1/52–9/30/53
			Geochemist	10/1/53–9/30/58
Faustite	$(Zn,Cu)Al_6(PO_4)(OH)_8 \cdot 5H_2O$	George Faust	Fellow	1933–1934
Fingerite	$Cu_{11}(VO_4)_6O_2$	L. W. Finger	Crystallographer	7/1/69–6/30/99
Fleischerite	$Pb_3Ge(SO_4)_2(OH) \cdot 3H_2O$	Michael Fleischer	Asst. Physical Chemist	3/1/36–12/31/38
Gaidonnayite	$Na_2ZrSi_3O_9 \cdot 2H_2O$	Gabrielle Donnay	Crystallographer	1/1/55–9/1/70
Greigite	Fe_3S_4	Joseph W. Greig	Petrologist	10/1/22–6/30/60
Haggertyite	$(Ba,K)(Ti_5Fe^{2+}_{\ 4}Fe^{3+}Mg)O_{19}$	Stephen E. Haggerty	Fellow	6/1/68–1/31/71
Hendricksite	$K(Zn,Mn)_3(Si_3Al)O_{10}(OH)_2$	Sterling B. Hendricks	Physicist	9/15/26–1/1/27
Henrymeyerite	$BaFe^{2+}Ti_7O_{16}$	Henry O. A. Meyer	Fellow	9/1/66–8/31/68

Mineral	Formula	Name	Position	Dates
Ingersonite	$Ca_3MnSb_4O_{14}$	H. Earl Ingerson	Asst. Physical Chemist	3/1/35–12/31/45
			Petrologist	1/1/45–3/15/47
Joesmithite	$PbCa_2(Mg,Fe^{2+},Fe^{3+})_5Si_6Be_2O_{22}$	Joseph V. Smith	Fellow	9/25/51–9/24/54
Kullerudite	$NiSe_2$	Gunnar Kullerud	Geochemist	9/7/54–12/31/70
Larsenite	$PbZnSiO_4$	Esper S. Larsen, Jr.	Asst. Petrologist	9/1/07–6/1/09
Lindsleyite	$(Ba,Sr)(Ti,Cr,Fe,Mg,Zr)_{21}O_{38}$	D. H. Lindsley	Petrologist	7/1/62–8/31/70
Londonite	$(Cs,K,Rb)Al_4Be_4(B,Be)_{12}O_{28}$	David London	Fellow	9/1/81–1/15/83
Loveringite	$(Ca,REE)(Ti,Fe^{3+},Cr)_{21}O_{38}$	John F. Lovering	Guest Investigator	3/1–5/1964
Mcbirneyite	$Cu_3(VO_4)_2$	Alexander R. McBirney	Research Associate	3/1/75–6/30/75
Merwinite	$Ca_3MgSi_2O_8$	Herbert E. Merwin	Assistant Petrologist	9/15/09–12/31/19
			Petrologist	1/1/20–3/1/45
			Research Associate	11/1/46–10/30/59
Niggliite	$PtSn$	P. Niggli	Research Associate	1913–1914
Paulmooreite	$Pb_2As_2^{3+}O_5$	Paul B. Moore	Guest Investigator	11/1/63–1/1964
Posnjakite	$Cu_4(SO_4)(OH)_6 \cdot H_2O$	Eugene Posnjak	Research Associate	10/1/12–1/14/13
			Assistant Chemist	1/15/13–6/30/15
			Chemist	7/1/15–9/30/46
Prewittite	$K_2Pb_3Cu_{12}Zn_2O_4(SeO_3)_4Cl_{20}$	Charles T. Prewitt	Director	7/1/86–6/30/98
			Crystallographer	7/1/98–present
Rambergite	$\delta\text{-}MnS$	Hans Ramberg	Research Associate	10/1/55–9/30/57
Ramdohrite	$Pb_6Ag_3Sb_{11}S_{24}$	Paul Ramdohr	Research Associate	11/1/60–4/30/61
			Guest Investigator	1/1/61–4/30/62
				11/1/64–6/30/64
Rankinite	$Ca_3Si_2O_7$	George A. Rankin	Asst. Physical Chemist	11/1/07–4/30/16
				1/1/20–6/30/20
Ringwoodite	$(Mg,Fe^{2+})_2SiO_4$	A. E. Ringwood	Fellow	10/1/63–9/30/64

(cont.)

Table 2.2 (*cont.*)

Mineral name	Formula	Staff	Position	Dates (m/d/y)
Roedderite	$(Na,K)_2(Mg,Fe^{2+})_5Si_{12}O_{30}$	Edwin W. Roedder	Fellow	9/1/47–8/30/48
Sahamalite – (Ce)	$(Mg,Fe^{2+})Ce_2(CO_3)_4$	Thure G. Sahama	Visiting Investigator	1/1/47–1/27/49
Schairerite	$Na_{21}(SO_4)_7 F_6Cl$	J. Frank Schairer	Physical Chemist	9/1/27–6/30/69
			Research Associate	7/1/69–9/26/70
Schreyerite	$V_2^{3+}Ti_3O_9$	Werner Schreyer	Fellow	1/1/58–12/31/61
			Staff Associate	1/1/62–12/31/63
Tilleyite	$Ca_5Si_2O_7(CO_3)_2$	Cecil E. Tilley	Guest Investigator	1/1956–7/1967 (6-month intervals)
Tombarthite – (Y)	$Y_4(Si,H_4)_4 O_{12-x}(OH)_{4+2x}$	T. F. W. Barth	Petrologist	10/1/29–3/1/36
Tunellite	$SrB_6O_9(OH)_2 \cdot 3H_2O$	George Tunell	Petrologist	9/1/25–6/30/47
Viaeneite	$(Fe,Pb)_4S_8O$	Willy A. Viaene	Fellow	4/1/70–2/28/71
Wonesite	$(Na,K)(Mg,Fe,Al)_6(Si,Al)_8O_{20}(OH,F)_4$	David Wones	Fellow	2/5/57–5/31/59
Yagiite	$(Na,K)_3Mg_4(Al,Mg)_6(Si,Al)_{24}O_{60}$	Kenzo Yagi	Guest Investigator	10/1960–1/1961
Yoderite	$Mg_2(Fe^{3+},Al)_6Si_4(O,OH)_{20}$	Hatten S. Yoder, Jr.	Petrologist	6/1/48–6/30/71
			Director	7/1/71–6/30/86
			Director Emeritus	7/1/86–2003
Ziesite	$\beta\text{-}Cu_2^{2+}V_2^{5+}O_7$	E. G. Zies	Research Assistant	4/1/13–6/30/15
			Chemist	7/1/15–9/30/49
			Research Associate	10/1/49–6/30/67

had to be devised involving extensive alterations to the microscope and new techniques for obtaining accurate optical constants of the phases, especially where solid solutions are formed. The leader in these efforts was F. E. Wright who, in 1910, reported using a Zeiss No. 1 C microscope for photomicrography as a base.[53] The changes he made included:

1. simultaneous rotation of the nicols about the optic axis of the microscope;

2. simple, dust-proof mechanical stage with spring loaded screws that obviates errors due to lost motion in the screws;

3. rotating and removable part containing the iris diaphragm and polarizer;

4. an Abbe condenser and a large nicol prism so the entire condenser system can remain in position when low-power objectives are used;

5. quartz plate below condenser to facilitate the determination of the ellipsoidal axis;

6. the Betrand lens mounted on a slide so that different magnifications of the interference figure can be obtained;

7. introduction of a second iris diaphragm below the ocular so interference figures can be obtained by the Lasaulx method without the Betrand lens; and

8. a new occular with accompanying plates for use in measuring birefringence, the optic axial angle, and extinction angles of minerals in powdered form or thin section.

To aid the exact measurement of the refractive index, a set of refractive index liquids were prepared ranging from 1.450 to 1.840 with each successive index differing only by 0.005. The liquids were calibrated with an Abbe-Pulfrich total refractometer and checked every three months. Liquids with indices of refraction up to 2.92 were eventually designed by Merwin and Larsen,[54] and with a new monochromatic light source very accurate indices of refraction were obtained.[55]

It is essential that every run product be examined optically; however, single crystal and powder X-ray diffraction studies were not only confirmatory, but critical especially for the very fine-grained products. The Debye-Scherrer technique (1916)[56] for the analysis of powdered specimens became available and appears to have been used at the Geophysical Laboratory after 1920 when X-ray generators were purchased (see Chapter 7 on X-ray crystallography). By 1950, a new Norelco Geiger counter focusing diffractometer was installed and powder X-ray patterns became routine.[57] An automatic slide changer,[58] on which twelve specimen slides could be mounted, greatly faciliated the analysis of the many run products on the several systems being investigated by J. F. Schairer, for example. Powder X-ray diffraction remains one of the most useful adjuncts to initial optical identifications.

3

PRESSURE

Introduction

Reproduction of the terrestrial conditions of temperature (T) and pressure (P) was a goal of experimenters from the outset. Because of the lack of knowledge of the mineral composition of the Earth's zones, their physical and thermal properties, the extent of lateral variation, and theoretical constraints, only a crude estimate of the P and T in the Earth could be made (Table 3.1). Nevertheless, the range of P and T in the Earth was evident. The advisors for the establishment of the Geophysical Laboratory (GL) emphasized the need for critical observations especially on the effect of pressure.

Those investigators who were originally employed by the US Geological Survey (USGS) had initiated programs involving pressure. In 1893, for example, Barus attained 2000 bars at 400 °C[1] while studying the effect of pressure on natural rocks at the temporary laboratory of the USGS established at the American Museum of Natural History in New York. Grants were given to Richards (1903) for compressibility measurements of liquids and gases, to F. D. Adams (1903) for rock deformation, and Becker (1904) for elasticity and plasticity of rocks. The latter studies were supported by Woodward, who had special interest in schistosity. The distinction between hydrostatic (uniform) pressure versus stress (non-uniform pressure) was clearly recognized. In 1907 Day stated that several pieces of apparatus had been designed and built for achieving high pressures, but their installation was postponed "until a special laboratory and more favorable conditions shall become available"[2] than "the crowded quarters" then occupied in an office building. The first high-pressure experiments in the new building were actually performed with apparatus brought in from Germany by Dr. Albert Ludwig.

Apparatus development

Ludwig was granted patents both in Germany in 1900 and in Great Britain in 1901 for a "vessel for the reception and heating of high-pressure gases." On application to Dr. Day, Ludwig was appointed a Research Associate at the Laboratory in 1908 and his apparatus transported to the USA.[3] He published results in 1909 in which he describes a "pressure bomb which held

Table 3.1 *Approximate P–T conditions at critical seismic discontinuities*

Zone	Depth (km)	P (kbar)	T (°C)
Crust	40	10	500
Upper mantle	410	140	1320
Transition zone	660	235	1500
Lower mantle	2900	1360	3700
Inner-outer core boundary	5150	3250	5500

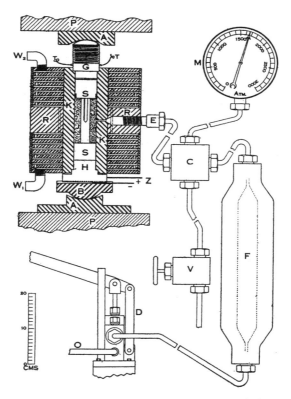

Figure 3.1 Apparatus of Johnston and Adams (1911) patterned after Ludwig (1909). It consists of a ring of nickel-steel (R) and a number of rings of boiler plate shrunk on a steel cylinder (K). The pressure medium is paraffin oil supplied by a pump (D) and measured with a Bourdon Gauge (M). The heater is an electric furnace (stippled) insulated with soapstone blocks (S). *Source*: From J. Johnston and L. H. Adams, "The influence of pressure on the melting points of certain metals," *Am. J. Sci.*, 4th series, 31 (1911), 503, figure 1.

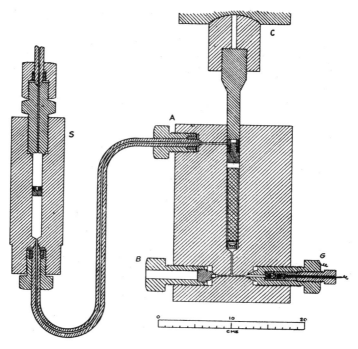

Figure 3.2   The apparatus of Adams, Williamson, and Johnston (1919) for compressibility measurements. The pressure vessel is of vanadium steel and mounted in a 500-ton hydraulic press (see Figure 3.3). The piston movement that measures volume change was read on an Ames dial micrometer. The pressure was measured by the change of electrical resistance of a manganin wire gauge (G). *Source*: From L. H. Adams, E. D. Williamson, and J. Johnston, "The determination of the compressibility of solids at high pressures," *J. Am. Chem. Soc.* 41 (1919), 13, figure 1.

tight at 700 °C and 17,000 atmospheres." The next major development was the vessel of Johnston and Adams[4] that appears to have been reconstructed from the one designed and built by Ludwig;[5] however, they only reported achieving 400 °C and 2000 atmospheres (Figure 3.1). An apparatus (Figure 3.2) was also constructed for determining cubic compressibility of various rocks, metals, and minerals in which kerosene was the pressure medium to 12,500 atmospheres at room temperature.[6] The volume measurement was critical in calculating pressures at depth. The same hydraulic press (Figure 3.3) was used to contain another design (Figure 3.4) in which Smyth and Adams[7] studied carbon dioxide to 1390 °C and 1000 atmospheres because of their interest in the carbonate rocks. It was presumably in this device that Eskola in 1922 first used sealed platinum tubes to hold olivine diabase and water at 850 °C and 2800 atmospheres.[8] He was particularly interested in

Figure 3.3 Hydraulic press, pressure vessel, micrometer and connections for the Adams, Williamson, and Johnston apparatus. *Source*: As Figure 3.2, p. 15, figure 3.

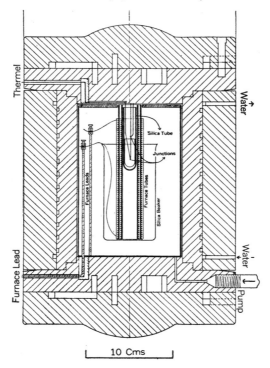

Figure 3.4 High-pressure apparatus of Smyth and Adams used for the study of carbon dioxide up to 1390 °C and 1000 atmospheres. From F. H. Smyth and L. H. Adams, "The system calcium oxide–carbon dioxide," *J. Am. Chem. Soc.* 45 (1923), 1172, figure 1. By permission of the American Chemical Society.

the serpentinization and alteration of basic rocks. In the 1930s Goranson, modified the Smyth and Adams design[9] thereby extending the range to 4000 atmospheres (see Chapter 4 on volatile components).

In 1921 Dr. Adams became interested in seismology (see Chapter 12 on geophysics) and well "appreciated that a knowledge of the elasticity of materials of the Earth's crust would enhance the value of seismological observations" (Adams' file). He continued, therefore, to measure the electric properties of diamond and the elastic properties of basic rocks and their mineral constituents.[10] He teamed up with J. W. Green at DTM to study the influence of hydrostatic pressure on the critical temperature of magnetization for iron and other materials. It became quite clear from the lack of effect of pressure up to 3600 atmospheres on the Curie point for five ferromagnetic materials, that the nickel–iron core of the Earth "has no important direct influence on the Earth's magnetic field."[11] Adams' interest in compressibility no doubt also stemmed from his appreciation of the fundamental role of volume in thermodynamic considerations[12] as well as a measure of the

Figure 3.5 Photograph of Tuttle's hot-seal hydrothermal apparatus in quenching position, with air jet around sample holder as modified by Van den Heurk. From J. Van den Heurk, "Improved hydrothermal quenching apparatus," *Bull. Geol. Soc. Am.* 64 (1953), 992, plate 1. By permission of the Geological Society of America.

deep-seated material outside the core. By 1940 R. W. Goranson and E. A. Johnson had reached 200,000 atmospheres in a two-stage cascaded pressure apparatus where sodium chloride was compressed to almost 30 percent.[13] With the appointment of Adams as director of the Geophysical Laboratory and the onset of World War II, high-pressure studies were temporarily laid aside.

After World War II there was a concerted effort to achieve higher pressures and temperatures using materials developed for other purposes during the war, and especially to investigate systems related to metamorphic rocks containing volatile components. The hot-seal apparatus (Figure 3.5) of O. F. Tuttle (1948),[14] modified in 1953 by Van den Heurk,[15] and the cold-seal pressure vessel (Figure 3.6), using steels developed for machine-gun barrel liners,[16] were reminiscent of the earliest devices built in France, but with the additional capability of fixing the pressure independently. From this point on advances came rapidly. In a cold-seal arrangement, 5 kbar was attained

Figure 3.6   One type of cold-seal hydrothermal pressure vessel. Furnace is raised by
N. L. Bowen (right) to show vertical pressure vessel in place. On the left is O. F.
Tuttle connecting a pipe to the pumping system (Archives, GL).

with the aid of an intensifier (Figure 3.7).[17] Then, in 1950, Yoder constructed
an internally-heated, gas-media apparatus (Figure 3.8) that sustained 1650 °C
and 10 kbar.[18] The interest in minerals and rocks alleged to have formed only
under high pressures, that is, at depth in the Earth, was revived. Success was
achieved in the garnet family and with many metamorphic rock assemblages
including eclogite. One significant failure was in the synthesis of jadeite
($NaAlSi_2O_6$).[19] It was therefore a matter of some surprise that details of the
incredible synthesis of jadeite and other minerals were announced in 1953
by L. Coes.[20] Coes was generous in opening his laboratory for a visit in
December 1953 to a group of nine high-pressure investigators from the Geophysical Laboratory, National Bureau of Standards, US Geological Survey,
Harvard University and Pennsylvania State University. He had produced in a
solid-media apparatus[21] (Figure 3.9) the minerals kyanite, pyrope, chloritoid,
staurolite, jadeite, and a host of others that many investigators had struggled
unsuccessfully to obtain. In his report to the Norton Co. (Worcester Massachusetts; unpublished notebook A-39) Coes recorded runs up to 45 kbar
and 1000 °C. The trick in achieving these conditions was the in-house availability of tungsten carbide for pistons and hot-pressed alumina for the pressure vessel. His solid-media device was the basis for the very successful design

Figure 3.7 Compact cold-seal, pressure-vessel bench employing an intensifier for operation to 5000 atmospheres. Gauge reads 75,000 pounds per square inch, the equivalent of 5000 atmospheres. The unit was assembled in 1949 by Yoder. The furnace of the middle unit is open to show the vertical pressure vessel and the high-pressure cone-in-cone seal below the furnace. From H. S. Yoder, Jr., "Experimental mineralogy: achievements and prospects," *Bull. Mineral.* 103 (1980), 15 figure 20. By permission of the *European Journal of Mineralogy*.

(Figure 3.10) of Boyd and England in the late 1950s, in which 50 kbar and 1750 °C could be maintained.[22] Hundreds of these devices are now in use around the world for investigations on mineralogical problems of the upper mantle.

About the same time F. Birch, E. C. Robertson, and S. P. Clark[23] made a successful device employing a tapered pressure cylinder and achieved 1400 °C and 27 kbar (Figure 3.11). Those limits were exceeded by H. T. Hall employing a new concept with a tetrahedral hydraulic ram assembly that reached 2000 °C and 100 kbar (Figure 3.12).[24] Hall is also credited with the belt apparatus[25] (Figure 3.13) in which the synthesis of diamond was reported in experiments up to 3000 °C and 100 kbar at the General Electric Research Laboratory. As these devices became more massive, another team at the US National Bureau of Standards changed to a miniature device based on Bridgman's anvil concept.[26] The sample was squeezed between two opposing diamond anvils and pressures up to 160 kbar and low temperatures (<175 °C) were sustained. Improvements in that patented design were made by H-k. Mao and P. M. Bell in the 1970s,[27] and led to 1.72 megabar at temperatures

Figure 3.8   Photograph of early internally-heated, gas-media apparatus of Yoder that sustained 1650 °C and 10,000 atmospheres. The external water jacket for cooling the pressure vessel and the ice bath for the thermocouple base point are visible. The gas intensifier and gauge for measuring piston travel are on the right (Archives, GL).

up to 3500 °C when an appropriate laser beam is applied to which the diamond is transparent (Figure 3.14).[28] Those conditions reproduce the pressures and temperatures believed to exist at the core-mantle boundary of the Earth. The conditions at the center of the Earth, 3.6 megabars and 5600 °C, have not as yet proved feasible;[29] however, pressures up to 5.5 megabars have been sustained.[30]

Pressure units and calibration

Measuring and calibrating temperature, pressure, defined as force per unit area, brought a new array of problems in measurement and calibration. The earliest pressure-measuring devices were the mercury column, Bourdon

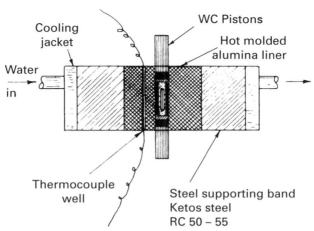

Figure 3.9 Cross-sectional drawing of Coes' solid-media, high-pressure apparatus using a hot-pressed Al_2O_3 core, supported by steel bands and tungsten carbide (WC) pistons. From Yoder, "Experimental mineralogy," *Bull. Mineral.* 103 (1980), 17, figure 23. By permission of the *European Journal of Mineralogy.*

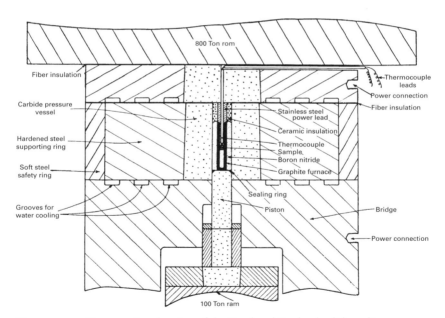

Figure 3.10 Cross-section drawing of the Boyd and England solid-media, high-pressure apparatus for use to 1750 °C and 50 kbar. From F. R. Boyd and J. L. England, "Apparatus for phase-equilibrium measurement at pressures up to 50 kb and temperatures up to 1750 °C," *J. Geophys. Res.* 65 (1960), 742, figure 1. By permission of the American Geophysical Union.

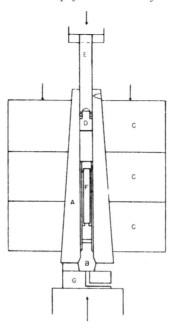

Figure 3.11   Cross-section of apparatus of Birch, Robertson, and Clark employing a tapered cylinder. From F. Birch, E. C. Robertson, and S. P. Clark, "Apparatus for pressures of 27,000 bars and temperatures over 1400 °C," *Ind. Eng. Chem.* 49 (1957), 1965–1966. By permission of American Chemical Society.

tube, and piston displacement. The displacement of a column of mercury could be used up to about 6 atmospheres and was expressed in mmHg. The deflection of an arcuate Bourdon tube sealed at one end was used mainly for liquids and gases up to 100,000 pounds per square inch (p.s.i.). That device was calibrated with the displacement of a piston loaded by weights. The piston had to be corrected for friction, which could be minimized by rotating the piston. Whereas those staff members with an engineering background used p.s.i., others preferred the physicists' units of kg/cm² or dynes/cm² (106 dynes/cm² = 1020 kg/cm² = 0.987 atm). In 1919 Adams, Williamson, and Johnston chose the unit megabar on the grounds that kg/cm² varied with the force of gravity, and, therefore, changed with geographical locality.[31] That unit persisted until about 1929, sometimes spelled megabarye, but by 1931 atmospheres had come back into vogue. In October 1960 the General Conference on Weights and Measures in Paris formulated a new International System of Units (SI), but it was not adopted in the USA until July 1971. Pressure was then to be measured as newtons/m², which became known in 1960 as the pascal[32] (1 bar = 105 pa = 0.987

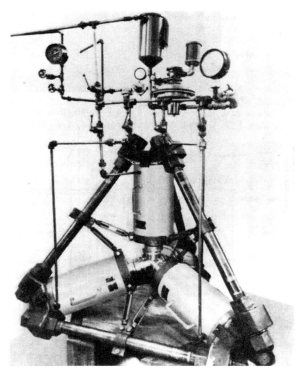

Figure 3.12 Tetrahedral hydraulic ram assembly of H. T. Hall (1958) developed because of delays imposed by government secrecy orders preventing disclosure of "belt" apparatus (see Figure 3.13). From H. S. Yoder, Jr., "Experimental Mineralogy: achievements and prospects," *Bull. Mineral.* 103 (1980), 18, figure 28. By permission of the *European Journal of Mineralogy*.

atm = 14.504 p.s.i.). This unit appears to be preferred by most journal editors and is beginning to dominate over the investigators' choice.

A major contribution to the measurement of pressure came in 1903 when E. Lisell found that the resistance of manganin wire changed almost linearly with pressure.[33] P. W. Bridgman also used the decrease in electrical resistance of mercury as a secondary gauge.[34] Nevertheless, he returned to the manganin gauge in the same year (1909), after discovering several disadvantages in using a liquid. The manganin gauge was adapted by Adams, Williamson, and Johnston in 1919[35] and was later developed into an instrument of high precision.[36]

When the pressures began to escalate, especially in the diamond-anvil, high-pressure cell, a rapid and convenient secondary pressure measurement was employed at the National Bureau of Standards. It is based on the pressure-induced shift in the wavelength of the R1 fluorescence peak from ruby.[37]

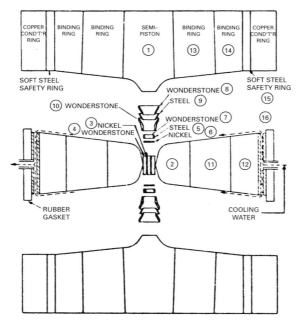

Figure 3.13   Cross-section of Hall's "belt" apparatus in which synthesis of diamond was achieved. From H. T. Hall, "Ultra-high-pressure, high-temperature apparatus: the 'belt'." *Rev. Sci. Instrum.* 31 (1960), 125, figure 1. By permission of American Institute of Physics.

Because the shift is also affected by temperature, a thermal correction factor is required. The gauge was extended at the Geophysical Laboratory to over a megabar by Mao and his colleagues in the late 1970s, who estimated the total random error to be ± 3 percent.[38] The wavelength shift of ruby was initially calibrated against X-ray diffraction data for NaCl to 20 GPa and subsequently extended to pressures above 100 GPa by calibration against shock compression of several metals. The stability of Al_2O_3 at high temperatures makes it ideal as an internal pressure sensor.

With improvments in spectroscopic techniques it is possible to measure the acoustic velocity with Brillouin scattering (V_Φ) and the density (ρ) with X-ray diffraction and to calculate the pressure from $P = V_\Phi^2 d\rho$. In the late 1990s this was considered a primary pressure standard and was used on olivine and ice and MgO to substantial pressures.[39]

Measuring structural phase transitions is the most common and conveniently employed technique for calibrating a pressure gauge. Those transitions used at the Geophysical Laboratory are included in Table 3.2. Some of the transitions are sluggish and care must be taken to carry out rate studies

Table 3.2 *Structural transitions for pressure gauge calibration*

Compound	Transition	Temperature	Pressure (kbar)
CCl_4	I = II	20 °C	3.3
Hg	L = I	0 °C	7.569 ± 0.002
H_2O	L = VI	25 °C	9.6
H_2O	VI = VII	25 °C	22.3
Bi	I = II	25 °C	25.38
Ba	I = II	25 °C	55 ± 2
Bi	III = V	25 °C	77±

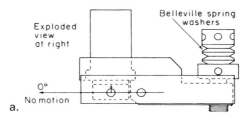

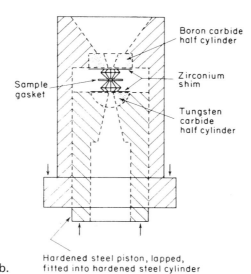

Figure 3.14 (a) Cross-section of modified design of diamond-anvil, high-pressure cell of Mao and Bell. (b) Detail of cell body with outside diameter of 3.5 cm. *Source*: From H.-k. Mao and P. M. Bell, "Design of a diamond-windowed, high-pressure cell for hydrostatic pressures in the range 1 bar to 0.5 Mbar," *CIW Year Book* 74 (1975), 403, figure 3.

and to determine the extent of hysteresis. Triple points are clearly superior to single transitions.

It was clear from the debates in 1968 at the "Symposium on Accurate Characterization of the High-Pressure Environment"[40] that accurate values could not as yet be determined. Observing the change in volume of a simple substance, such as NaCl, Au, or MgO, provided a range of pressure if the temperature could be ascertained. Here again, the effect of pressure on the thermocouple electromotive force was subject to debate. The *P* effect on chromel/alumel and platinum/platinum 10 rhodium had been measured by Bridgman in 1918,[41] Birch in 1939[42] and more recently by Bell, England, and Boyd in 1971[43] up to 590 °C and 5 kbar. Aside from the thermal gradients in high-pressure apparatus, the accuracy of both pressure and temperature measurements are uncertain.

4

VOLATILE COMPONENTS

Hydrothermal systems

The Advisory Committee and Committee of Eight both recognized the need to study the effect of volatiles on rock formation and alteration. They specifically pointed out the need for investigations into gas solubility in magmas, hydrous mineral stability, rock–water interaction, and hydration and dehydration. The first hydrothermal studies at the Geophysical Laboratory (GL) were carried out by E. T. Allen and J. K. Clement in 1908 on the role of water in tremolite.[1] They used a "bomb" (Figure 4.1), carried over from the US Geological Survey by Allen, in which the pressure was calculated from the degree of filling. They suggested that tremolite contained "dissolved water" as a solid solution that could be slowly driven off with increasing temperature and, conversely, restored under pressure. This view was clarified some twenty years later by E. Posnjak and N. L. Bowen who showed that water in tremolite entered the molecule in definite proportions in accord with W. T. Schaller's calculation of the chemical composition and the structural determination of B. E. Warren.[2] The same bomb was used in a study of K_2SiO_3-SiO_2-H_2O by G. W. Morey and C. N. Fenner in 1917[3] who were no doubt stimulated by the analyses of volcanic gases that had been collected at Kilauea, Hawaii by A. L. Day and E. S. Shepherd in 1912. Morey and Fenner's analyses revealed that "water is usually if not always the chief agent in volcanic activity."[4] Their pioneering study (Figure 4.2) revealed many of the principles now well established in igneous petrology. They showed that:

1. mineral solubility in solution increases with temperature;
2. pressure increases from the crystallization of non-volatile phases;
3. water is an original component of the magma;
4. water has potency in lowering melting points of silicates;
5. water has enormous influence in lowering viscosity of a magma; and
6. highly compressed vapor at high temperature readily transports materials.

They recorded results at temperatures of 600 °C and pressures of 340 atm, calculated from the van der Waal's equation, the first time a solvent had been studied above the critical conditions (374 °C and 217.5 atm) for H_2O.

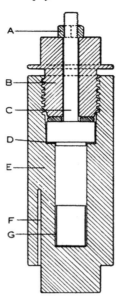

Figure 4.1   Construction of bomb used by Allen and Clement. Pressure is calculated from the degree of filling. The vessel is cooled quickly by plunging into cold water. *Source*: From G. W. Morey and C. N. Fenner, "The ternary system $H_2O-K_2SiO_3-SiO_2$," *J. Am. Chem. Soc.* 39 (1917), 1181, figure 1.

The volcanic gas sampled at Kilauea, Hawaii, also contained substantial amounts of CO_2. F. H. Smyth and L. H. Adams studied the decomposition of calcite[5] in their apparatus (see Figure 3.4) that sustained 1400 °C and 1 kbar; however, the first successful use of internal heating appears to have been by H. E. Boeke in 1912,[6] also for observing the decomposition of calcite.

In order to investigate a variety of gases that may be reactive with an internal heating element of the furnace, P. Eskola sealed the volatile in a platinum tube. His first experiments used the Smyth and Adams apparatus in 1922, when he was a visiting investigator at the Laboratory, in which an olivine diabase and water were held at 850 °C and 2.8 kbar.[7]

In 1931 Greig, Shepherd, and Merwin[8] made the incredible discovery that granite, freed from volatiles, melted at a lower temperature than basalt, contrary to popular opinion. The first systematic results on the melt curves of rock-forming minerals and a rock were obtained by R. W. Goranson[9] using a modified Smyth and Adams design. According to Goranson's results on a granite from Stone Mountain, Georgia, the water content of the melt was 9.35 wt. percent at 4 kbar, and 900 °C. The *P–T* curves for albite and orthoclase, major constituents of the granite in the presence of H_2O, were given in his results of 1938[10] (Figures 4.3a and 4.3b). He emphasized the conclusions of Morey and Fenner regarding the great lowering of the melting temperatures

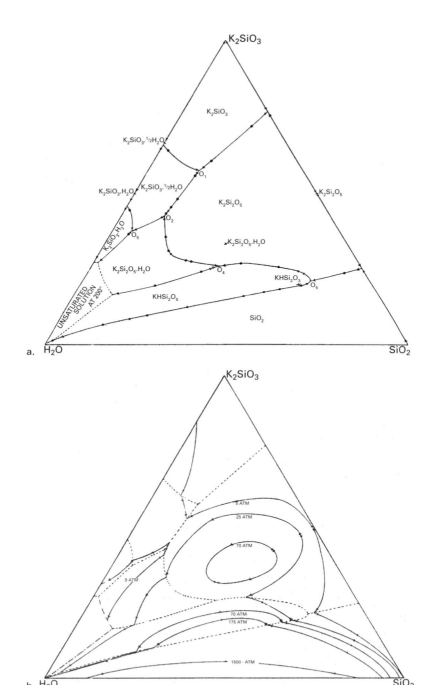

Figure 4.2 (a) Diagram of K_2SiO_3-SiO_2-H_2O showing determined boundary curves of phases. Arrows show direction of falling temperature. (b) Isobaric polythermal saturation curves. Arrows show direction of falling temperatures. Curves illustrate effect of varying water pressure. *Source*: As Figure 4.1, p. 1211, figure 8, and p. 1223, figure 13.

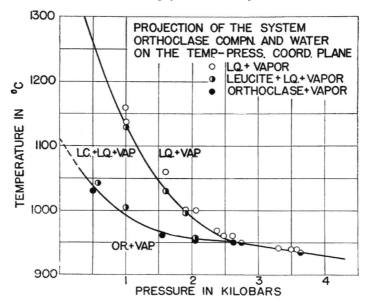

Figure 4.3a Projection of freezing-point curve for orthoclase-water illustrating the incongruent melting below 2.6 kbar and congruent melting above that pressure. From R. W. Goranson, "Silicate-water systems," *Am. J. Sci.,* Day Volume 35-A (1938), 89, figure 5. Reprinted by permission of the *American Journal of Science.*

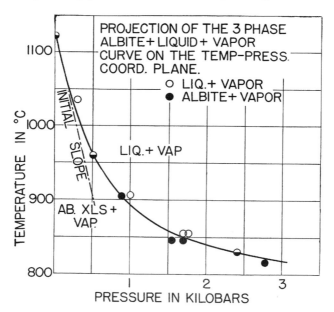

Figure 4.3b Projection of freezing-point curve for albite-water. *Source*: From R. W. Goranson, "Silicate -water systems, " *Am. J. Sci.*, Day Volume 35-A (1938), 84, figure 3. Reprinted by permission of the *American Journal of Science.*

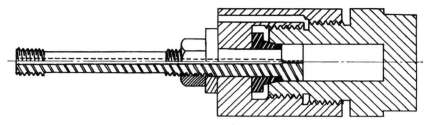

Figure 4.4 The Morey bomb. Built in 1941 for hydrothermal synthesis at 500 °C and 3 kbar. Outside diameter = 6.4 cm. From G. W. Morey, "Hydrothermal synthesis," *J. Am. Ceram. Soc.* 36 (1953), figure 2. Reprinted with permission of the American Ceramic Society.

with water and the substantial increase in pressure on cooling of the magma. It should be noted that in 1937 Morey and E. Ingerson, after a very detailed review of the literature, concluded that "practically no reliable quantitative information" was available on the role of water.[11]

By 1941, a way was found in externally heated vessels to fix the pressure independently from the degree of filling.[12] The "Morey bomb" (Figure 4.4) could be attached directly to a pump and gauge. Morey carried out hydrothermal mineral synthesis at 500 °C and 3 kbar for long periods of time. His results included the solubility of quartz in superheated steam. The Morey bomb was convenient to use and became a standard apparatus around the world. With a temperature regulator and a photo-electric cell that responded to changes in the pressure, the pressure could be regulated by means of an electric pump.

After World War II, the hot-seal apparatus of O. F. Tuttle (see Figure 3.5), the cold-seal apparatus (see Figures 3.6 and 3.7) and the internally heated, gas media apparatus of Yoder (Figure 3.8) were focused on hydrothermal systems.[13] The experimental study of metamorphic rock systems was initiated with studies of $MgO-SiO_2-H_2O$ (Bowen and Tuttle)[14] and $MgO-Al_2O_3-SiO_2-H_2O$ (Yoder).[15] Some of the well-established isograd reactions were determined, the concept of a water-deficient region recognized, reversals in isograd sequence due to differences in water pressure established, a new interpretation of retrograde metamorphism offered, and the stability of specific metamorphic minerals outlined. Subsequently, common mineral groups in metamorphic rocks were studied experimentally. These included the micas, chlorites, cordierite, serpentine, talc, zeolites, and garnets.

In light of Goranson's early success in dealing with a natural granite,[16] Yoder and C. E. Tilley undertook, in 1962, a study of natural basalt, the most common rock in the crust.[17] The knowledge gained from $An-Di-Fo-H_2O$ and the behavior of the three major phases in basalt as well as albite, nepheline, and quartz in the presence of water provided guidance.[18] In addition, the key amphibole end members had been studied a few years earlier by F. R. Boyd and

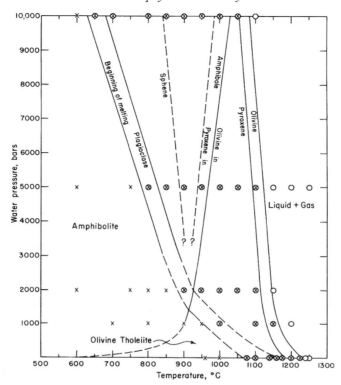

Figure 4.5 Projection of natural olivine tholeiite-water system from data on 1921 lava, Kilauea, Hawaii, demonstrating conversion of basalt to amphibolite. From H. S. Yoder, Jr. and C. E. Tilley, "Origin of basalt magmas: an experimental study of natural and synthetic rock systems," *J. Petrol.* 3 (1962), 449, figure 27. By permission of Oxford University Press.

W. G. Ernst.[19] On this basis, Yoder and Tilley undertook their hydrothermal investigation of the multicomponent basalt system with water. Three natural basalts were converted into amphibolites and their melting relations outlined (Figure 4.5). The results were applicable to submarine volcanism, water content of source magma, metamorphism of basalts, the proposed amphibolite layer in the crust, formation of anorthosites, reaction series concept, and the interrelation of basalts and granites. In spite of the firm basis developed from pure systems, some geologists were reluctant to accept the results on the "dirty" natural systems. For most geologists, the similarity of results were held in awe and applied without question.

Metasomatism

The presence of a static solvent can explain some mineral textures that result from metamorphic isograd reactions. Other textures involve the transport of

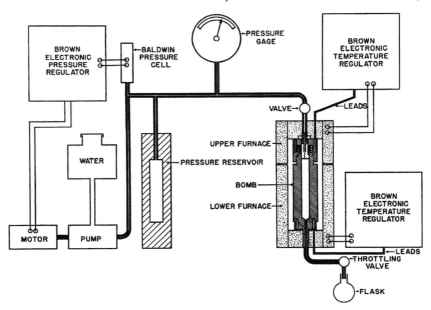

Figure 4.6 Diagrammatic outline of flow-through, high-pressure apparatus. Distilled water at constant pressure is passed over the sample in the bomb held at a fixed temperature. The exiting fluid is condensed, weighed, and analyzed. From G. W. Morey and J. M. Hesselgesser, "The solubility of some minerals in superheated steam at high pressures," *Econ. Geol.* 46 (1951), 822, figure 1. Reprinted with permission of *Economic Geology*.

material by the flow of a solvent in response to gradients in fluid pressure. The isograd reactions are usually demonstrated by sealed-tube experiments wherein the fluid is static.

A flow-through system was set up by Morey and J. M. Hesselgesser in 1951 for the purpose of obtaining the solubility of a wide range of minerals in superheated steam at high pressures (Figure 4.6).[20] Their focus was on the genesis of pneumatolytic and vein deposits. Particular attention was paid to solubility of a mineral versus decomposition as observed in the case of enstatite. A flow-through system was also set up by T. R. Filley in 1998 to investigate organic systems involving dilute formic acid, FeS, and water. The flows of up to 15 μL/min at pressures up to 70 MPa and $T = 220\ °C$ were possible. A computer was used to control the pressure or flow rate from the syringe pumps. Mixing fluids of different composition could thereby be investigated in titanium reactor vessels.

It is often assumed in geological problems that the water pressure acting on a mineral at depth in the earth is equal to the force of the overlying rock. The laboratory experiments concerning mineral stability, where the water pressure is the total pressure on the system, are also run on this assumption. In fact, however, the water pressure may be slightly greater if the strength

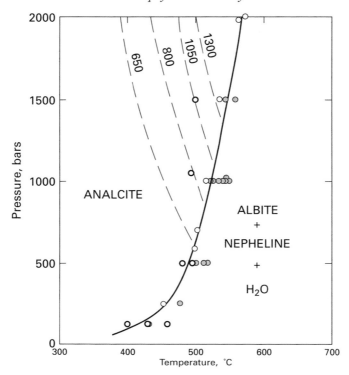

Figure 4.7 Stability diagram (solid line) for analcite ⇔ albite + nepheline H_2O
for $P_{H_2O} = P_T$. Limits of stability of analcite under fixed pressures of argon
(dashed curves) are deduced from Greenwood data for the 500 °C isotherm.
Source: From H. J. Greenwood, "Water pressure and total pressure in metamorphic
rocks," *CIW Year Book* 59 (1960), 61, figure 12.

of the confining rock is taken into account, or may be less, depending on
the accessibility of the surface to the water. In many geological environments
water has access to the surface by means of grain boundary diffusion or by flow
through fractures. For this reason it is necessary to consider mineral stability
under water pressures different from those of the total pressure acting on the
system.

The first attempt to evaluate those conditions was undertaken by H. S.
Yoder, Jr., in the early 1950s, with the system Analcite-H_2O-Argon.[21] According to theory, the upper stability limit of analcite would be reduced as the
total pressure was increased. The lowering of the equilibrium temperature
was not observed by him because the argon was introduced in the same pipe
and acted as a piston to increase the water pressure. By mixing the gases
in advance, H. J. Greenwood was later able to show the lowering of the
stability of analcite where $P_{H_2O} < P_{total}$ (Figure 4.7).[22] In addition, Greenwood showed that the proportion of two volatile components, for example

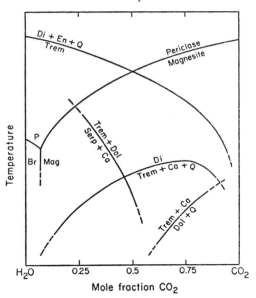

Figure 4.8 Metamorphic reactions involving two volatile components, H_2O and CO_2, illustrating the influence of proportions on the thermal stability of the assemblages. *Source*: From H. J. Greenwood, "Metamorphic reactions involving two volatile components," *CIW Year Book* 61 (1962), 84, figure 22.

H_2O and CO_2, have a strong influence on the stability region of a mineral assemblage (Figure 4.8).[23]

The influence of H_2O and CO_2 on melting was investigated in 1975 by M. Rosenhauer and D. H. Eggler, who demonstrated the importance of the ratio of volatiles is in the melting of albite.[24] Because of the slight solubility of CO_2 in albite, they found a temperature minimum trough on the divariant melting surface at about 5 kbar.[25] The special implication of those results is that a granitic magma with more than 50% of the H_2O–CO_2 fluid as CO_2 would melt along curves with a positive slope, whereas the melting curve of diopside would be negative at all ratios of CO_2 and H_2O (Figure 4.9). A magma rising in the crust would, therefore, have a variety of crystallization paths, depending on the volatile content.

The theory of infiltration metasomatism was laid out by D. S. Korzhinskii, J. B. Thompson and A. Hofmann.[26] The 1959 data of J. J. Hemley for the system K_2O-Al_2O_3-SiO_2-H_2O-HCl illustrated the potential of hydrothermal solutions to alter its wall rock as it ascended from higher temperature environments.[27] From thermodynamic analysis of the Hemley data, J. D. Frantz and A. Weisbrod showed that infiltration metasomatism depended mainly on fluid pressure gradients that may remain independent of the thickness of the reaction zones.[28] They were concerned with the volume

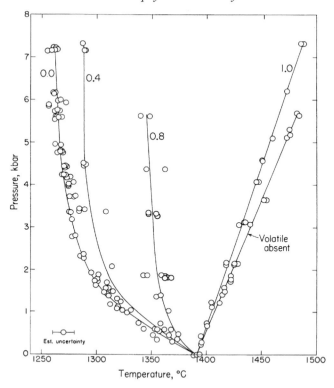

Figure 4.9 Melting of diopside, $CaMgSi_2O_6$, in the presence of CO_2 and H_2O and in the absence of volatiles. The numbers indicate $CO_2/(CO_2 + H_2O)$ of the vapor. *Source*: From M. Rosenhauer and D. H. Eggler, "Solution of H_2O and CO_2 in diopside melt," *CIW Year Book* 74 (1975), 475, figure 47.

changes as the reactions evolved, the changes in porosity, and the effects of the overburden pressure on the fluid pressure. Because most metasomatic processes involve two or more aqueous solutes, Frantz and Mao prepared a mathematical model for $CaO-MgO-SiO_2-H_2O-CO_2$. Qualitative rules were developed for the relative direction of movement of boundaries occurring by diffusion, infiltration, or both. The importance of the internal and externally imposed volume changes were again emphasized.[29] A year later, in 1976, their calculations illustrated that the thickness and modes of the zones of metamorphic reactions were unique functions of the mobilities of the different species and the topology of the solution phase diagram.[30] The beauty of their scheme was that it could be applied to any number of components in complex metasomatic deposits, such as porphyry copper. It appears that there is now a framework available on which to build useful experimental results from both static, closed systems as well as flow-through open systems.

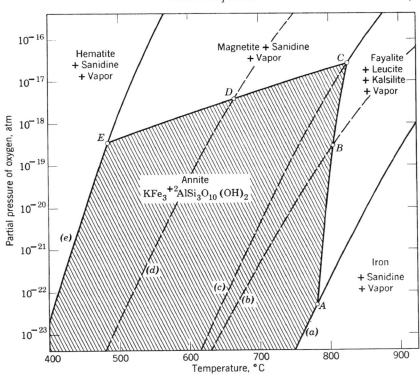

Figure 4.10 Isobaric P_{O_2}–T section for $P_{Total} = 2$ kbar illustrating the phase boundaries for annite. The buffers employed, a–e, are listed in the text. *Source*: From H. P. Eugster, "Reduction and oxidation in metamorphism." In P. H. Abelson (ed.), *Researches in Geochemistry* (John Wiley & Sons, 1959), p. 413, figure 9.

Redox reactions

Water undergoes thermal dissociation so that the species in a fluid are H_2O, O_2 and H_2: $H_2O \Leftrightarrow H_2 + \frac{1}{2}O_2$. The addition of water to a dry iron-rich assemblage would, therefore, change the oxidation state, possibly form hydrous phases, and drastically increase the diffusion rates. A very ingenious method was invented by H. P. Eugster in the late 1950s[31] to control the extent of oxidation in attempts to make annite, $KFe_3AlSi_3O_{10}(OH)_2$. He first surrounded the platinum capsule holding the appropriate constituents for annite with Zn + HCl to create a back pressure of hydrogen. Next Fe + HCl was used, then Fe + H_2O, and finally a large variety of assemblages that provided a source of hydrogen at a wide range of partial pressures. These assemblages were called buffers. Those now commonly used, as in the annite study (Figure 4.10), are:

(a)   fayalite, iron, quartz
(b)   wüstite, iron
(c)   magnetite, wüstite
(d)   fayalite, magnetite
(e)   hematite, magnetite.

The buffering technique opened the door to the study of a wide range of iron-bearing minerals.

In 1963 H. R. Shaw[32] developed another scheme for fixing the hydrogen partial pressure independently. In the conventional Shaw apparatus, a platinum tube filled with a solid support medium (e.g. coarse, crushed quartz) and welded closed at one end is soldered to a stainless-steel, high-pressure capillary and inserted in a hydrothermal pressure vessel. A controlled pressure of hydrogen is imposed on the platinum tubing through the capillary, and because the platinum is permeable to hydrogen, the platinum tubing acts as a semipermeable membrane through which the partial pressure of hydrogen in the vessel is controlled. Experimental samples welded in capsules permeable to hydrogen (e.g. Pt or Ag/Pd alloys) are then placed in the pressure vessel next to the platinum membrane. The technique is used to control the fugacity of hydrogen within the experimental capsule by controlling the partial pressure of hydrogen in the pressure vessel.

This apparatus was redesigned by Frantz and colleagues in the mid-1970s so that the sample was enclosed in a gold tube, relatively impervious to hydrogen.[33] This surrounded the platinum tube through which the hydrogen flowed under a controlled pressure. This setup greatly reduced the time for equilibration required for a conventional Shaw apparatus and provided for experimentation at any P_{H_2} rather than those fixed by the buffers in the Eugster system.

Diffusion metasomatism

Diffusion metasomatism depends in part on pore water, adsorbed water, interlayer water, and possibly structural water. Some of the observations such as porphyroblasts, hollow crystals, overgrowths, exsolution, zoning, morphological variations, amygdule fillings, and grain size variations clearly indicate that the rates of diffusion of materials is complex. The control is predominantly the rate at which materials are brought to or away from a reaction site whether crystalline or liquid. In 1921, N. L. Bowen was the first to evaluate the role of diffusion in accounting for the diversity of magmas.[34] He placed a diopside glass at the bottom of a crucible with a layer of plagioclase (Ab_2An_1) glass above. At a temperature above the melting of both layers, the crucible was held for definite periods, quenched, and the index of refraction of the glass measured at various depths. The diopside component diffused into the plagioclase layer with a k = 0.015 cm²/day. Bowen concluded that

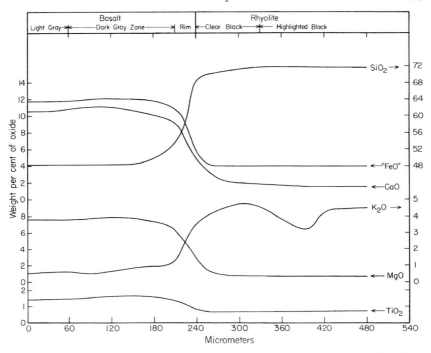

Figure 4.11 Compositional changes across the melted basalt-rhyolite interface for powdered samples placed end-to-end in a platinum tube and held at 1200 °C, 1 hr, at $P_{H_2O} = 1$ kbar. *Source*: From H. S. Yoder, Jr., "Diffusion between magmas of contrasting composition," *CIW Year Book* 70 (1971), 107, figure 2.

diffusion was a very slow process, inversely proportional to viscosity, and was not responsible for the border phases on large igneous bodies because of the high diffusivity of heat. He was well aware that other processes such as the circulation of solutions and convection would greatly overshadow diffusion alone.

In view of the fact that two magmas of highly contrasting composition could be generated from a single parent,[35] the question arose as to how these magmas once generated could be stored or retain their identity. Layers of a natural basalt and rhyolite were used by Yoder in a run at $P_{H_2O} = 1$ kbar, 1200 °C, and one hour. A sharp contrast was observed visually; however, analyses with an election microprobe revealed a diffusion zone of at least 300 μm (Figure 4.11). The diffusivity appears to have been dependent in part on the concentration of each element. He pointed out that immiscibility was *not* involved, because when powdered and thoroughly mixed the basalt and rhyolite were readily homogenized.[36]

The systems involving the volatile sulfur are discussed in Chapter 9 on ore geochemistry.

5

VOLCANOLOGY

The preliminary scientific program for the Geophysical Laboratory, as authorized by the Trustees in 1905 and as outlined by George F. Becker, emphasized the need for research on the "operative causes of volcanism." The Trustees were very receptive to a study of volcanology because, only five months before the formation of the Carnegie Institute of Washington (CIW) itself in 1902, Mont Pelée erupted, destroying the city and killing 28,000 inhabitants of St. Pierre, Martinique.[1]

In addition, H. S. Washington who served as an advisor to CIW was already working on the active volcanic regions of Italy. He held Grant No. 95 for three years, which funded his visits to the volcanoes of the eastern Mediterranean and his analytical work on the specimens collected, carried out in his private laboratory in Locust, New Jersey. That work culminated in 1906 with the publication of *The Roman Comagmatic Region*.[2] Fifteen rocks from Italy described by Washington were sufficiently unique to warrant new names.[3] The predominance of leucite- and sanidine-bearing rocks from this region is noteworthy.

Some of the volcanic regions studied by Laboratory staff will now be briefly described: Italy, Hawaii, California, Alaska, Wyoming, Dutch East Indies, Iceland, Mexico, Guatemala, Pantelleria. These studies gave rise to the need to understand explosive volcanism and experiments were directed to this problem.

Italy

Vesuvius was observed continually from 1903 to 1921 by F. A. Perret,[4] who paid special attention to the explosive stage of 1906, the repose period of 1906–1913, the subsidence events of 1910–1913, and the renewed eruptive period of 1914–1921. His classic monograph, published in 1924 by the CIW, contained supplementary photographs by Day and chemical analyses of some of the lavas by Washington.[5] Perret's dramatic "narrative" of the great eruption of April 1906 immerses the reader in an extensive variety of phenomena: continuous earthquakes, roaring sound, swirling clouds, electrical discharges, flashing arcs, rain of pisolites, hot avalanches scoring the cone, internal avalanches, and heavy ash falls containing "countless living caterpillars." He attributed the explosive phase directly to the gas content of the magma and the subsidence of the crater floor as a product of gas-fluxing and engulfment of the plug at the upper end of the magma column.

In the summer of 1914, Washington made a descent into the crater and described the various fumeroles, their colored accumulations of salts, and especially the variety of gases emitted (HCl, SO_2, SO_3, CO_2, and H_2S).[6]

Hawaii

In 1911 the Hawaiian Volcano Observatory was founded by T. A. Jaggar (MIT) in collaboration with R. A. Daly (Harvard), the Volcano Research Association of Hawaii, and the Geophysical Laboratory.[7] The measurement of the temperature of the lava lake in Kilauea and analysis of the role of gas in the flowing lavas were undertaken by E. S. Shepherd and F. A. Perret that year. These epoch-making studies involved the collection, via a cable across the lava lake, of an iron-bucket dip sample of the lava and immersion of a thermocouple pipe in the bubbling lake itself. In the summers of 1911 and 1912, Day and Shepherd collected and analyzed the gases in the active part of Halemaumau crater of Kiluaea. These early observations at Kilauea showed the close connection that existed between the quantity of gas being given off from the lava lake and the temperature and general activity. Efforts were made to collect gases fresh from the lake and uncontaminated with air, a task by no means easy. The unexpectedly large amounts of water found within the gases collected in 1912 was caused by condensation in the pipes that led the gases out to the collecting-tube. This interfered with the determination of the quantitative composition of the gases as a whole. The composition of the "fixed" gases could be determined, but not their relation to the water vapor. In 1917 gas samples were obtained in which the relation of water to the other constituents could be determined.[8]

Day and Shepherd clearly demonstrated that water was an original component of the lava, contrary to the prevailing view of the nonaqueous quality of magmatic gases.[9] Furthermore, they attributed the loss of gas as the reason for the structural change from the smooth, undulating Pahoehoe lava to the rough, blocky Aa lava.

California

Day and Allen next turned to Lassen Peak, California, after its catastrophic outbreak on May 30, 1914, the first eruption of a volcano within the continental boundaries of the USA in living memory (Figure 5.1).[10] In addition to their description of the eruptive activity, Day and Allen focused on the types of hot springs and fumeroles with field and laboratory measurements. The change from springs of acid character transporting pyrite to those of alkaline

Figure 5.1   Successive views of the explosion of June 14, 1914 of Mount Lassen, California, taken by B. F. Loomis, professional photographer. From Day and Allen, *Volcanic Activity and Hot Springs of Lassen Peak*, plate 1.

character was attributed to the interaction of the hot water with the silicate rocks. They concluded from a study of the conduit lava that the eruption was a "low-temperature phenomenon." The 1915 lava did not seem to them to be a new magma erupted through an old conduit. "It seems rather more like an old conduit lining, which had been fissured and reheated by a fresh influx of juvenile gases from below until finally it acquired sufficient mobility to allow it to be forced upward by pressure. The eruption of both pumiceous and dense lava together favors this hypothesis of irregular reheating along cracks in part of the old conduit lining."[11]

In the summer of 1921 drilling at "The Geysers"[12] in Sonoma County, California, was undertaken by J. D. Grant with the intent of utilizing the steam from fumeroles for power. Allen and Day[13] realized the holes would provide them with an opportunity to measure the temperatures and pressures with depth in a volcanic area. The fumeroles were rectilinear and assumed to be connected by faults. Whereas the surface temperatures were under the boiling point, the temperature increased to about 200 °C at a depth of about 500 feet where the pressure was 5 to 13 bars. The steam was primarily water with only 0.75–1.95% uncondensed gases dominated by CO_2 with much lesser amounts of CH_4, H_2, N_2, and H_2S. From these exceptional and rare data,[14] they concluded that the temperature gradient exceeded the Earth's gradient and was probably the result of a volcanic body at depth. One drill hole in the regional sediments and metamorphic rocks did penetrate gabbro. The heat, in their view, was from proximity to an intrusive body, and they predicted that the nature of the volcanic body would eventually be determined by the composition of the gases being released.

Alaska

In the meantime, Mount Katmai in Alaska had erupted in June 1912,[15] but it was 1916 before an expedition, organized by R. F. Griggs and supported by the National Geographic Society, reached the area. On that and subsequent expeditions to the Valley of Ten Thousand Smokes were C. N. Fenner, E. G. Zies, and E. T. Allen, who collected rocks, fumerole encrustations, measured the temperature of the hot springs, aspirated exhalations for the "insoluble" gases, and helped in the geologic mapping. They concluded that the vast sheet of siliceous rocks was not lava but was of pyroclastic origin, ejected as rhyolitic pumice through the fractured valley floor. The fumeroles (100–650 °C) were, therefore, of deep-seated origin and decreased in temperature with time. Through successive observations, the mineralogy of the encrustations changed as the temperature dropped, and because many minerals of economic value were formed, a relationship of ore deposits to volcanic exhalations was established. The analyses of the gases collected showed the highest contents of HCl and HF that had ever been detected (Figures 5.2 and 5.3). The hybrid nature of the rocks (also found at Lassen Peak) led Fenner to believe that a superheated rhyolite magma had melted fragments of old andesitic lavas and incorporated them into the erupted pumice and ash. The detailed analytical work, tied closely to the mineralogy and geology, established the value of a multidisciplinary approach to geologic problems.

Wyoming

The remarkable hydrothermal activity of Yellowstone National Park was described in another classic study by Allen and Day in 1935,[16] in which they recorded the physical and chemical changes of the fumeroles, geysers and thermal springs over a period of seven years. They defined the relationship of hot springs to fumeroles and contrasted the differences in rock alteration from the acid, mixed, and alkaline types. They identified superheated water up to 138 °C, which gave rise to violent effervescence, and they contributed to an understanding of discharge and its relation to rainfall.

Figure 5.2 Apparatus for analysis of gases. Removal of nitrogen from the inert gases was accomplished by furnaces in foregound. Photograph by Harper Snapp. *Source*: As Figure 5.1, p. 127, figure 69.

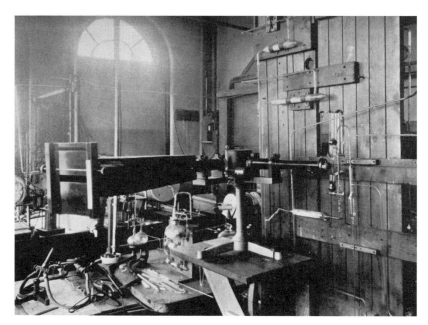

Figure 5.3 Spectrograph for analysis of inert gases used by Day and Allen. *Source*: As Figure 5.1, p. 130, figure 71.

The "lavas of abnormal appearance" described by Iddings (1899)[17] at Gardner River, Yellowstone Park, Wyoming, contain basalt xenoliths in rhyolite. These were interpreted by Fenner some thirty years later as the result of rhyolite magma penetrating an old basalt flow in complex networks of veins and dike-like bodies.[18] Apparently the corrosive action of the rhyolite greatly altered the basalt, removing some constituents in such a manner as to generate altered rocks on straight composition lines between basalt and rhyolite. The interpretation was questioned by Wilcox in 1944 who said the features were better explained by a mixing of nearly contemporaneous rhyolite and basaltic lavas whereby the cooler rhyolite was chilled against country rock, but may have interacted with a "liquid or mushy basalt."[19] Fenner responded,[20] holding to his previous view, that the basalt was much older and was penetrated by a low-density rhyolite of high fluidity. He also cited the circumstances at Katmai where basic rocks became softened and disintegrated by a rhyolite magma in contrast to Wilcox's idea that a rhyolite magma could chill and solidify a basaltic magma upon contact. Because of the topographic relationships, penetrations of rhyolite into the basalt, dispersion of rhyolite minerals in the basalt, and the relative thermal character of the two rock types, Fenner clearly recorded his disagreement with Wilcox.

Another major work on volcanic activity, supported in part by the Geophysical Laboratory, was F. R. Boyd's study of the welded tuffs and flows in the rhyolite plateau of the Yellowstone National Park, Wyoming.[21] Boyd mapped the plateau, determined its stratigraphy, and most importantly, discovered the Yellowstone caldera. His thermodynamic analysis combined with experimental evidence showed that tuffs can have temperatures of emplacement sufficiently high for them to weld. This analysis is of great importance to the understanding of welded tuffs.

Dutch East Indies

The Geophysical Laboratory participated in two expeditions in 1928 to the islands of Java and Bali in present-day Indonesia. The intent was to use a portable spectrograph to analyze the volcanic gases, but no favorable opportunities appeared. Instead, incrustations and lavas were collected and visual observations on the blue flames from Mount Raoengu, east Java, were made in spite of the most difficult terrain and turbulence in the associated steam vents. Tellurium and arsenic sulfide were present in the emanations at Papandajan, which altered the wall rocks on a high scale. Other rim volcanoes exhibited emanations of both sulfur and hydrochloric acid. In the Dieng area of Java an explosion took place early in 1928 that resulted in local roads being closed because of the high concentrations of CO_2 and H_2S.

Iceland

A further contribution to the study of hot springs and geysers was made by T. F. W. Barth who carried out the laboratory study of samples collected in Iceland.[22] The work was undertaken during the summers of 1934 and 1937, but publication was held up for eight years while Barth was detained in occupied Norway during World War II.

The siliceous volcanic rocks of Iceland were studied by K. Muehlenbachs in the 1970s to explain the uniquely low yet wide ranging ^{18}O contents.[23] At first, mixing of acidic and basaltic magmas was considered as a possible cause. Then interaction with meteoric waters also low in ^{18}O was considered, but the anhydrous nature of the Icelandic rhyolites was not compatible with that idea. Nevertheless, a two-stage model was presented by Muehlenbachs in which the partial melting of hydrothermally altered basalts yielded acidic rocks with the low but highly variable ^{18}O contents.

Mexico

The temperature of the Aguan flow of basaltic lava that issued during December 1944 from the Parícutin volcano in Mexico was estimated to be 1200 °C[24] and thus lies in the same range as that of other basaltic lavas. This value was obtained by means of an adequately protected chromel-alumel thermocouple connected to a portable potentiometer. Zies also presented evidence that the gases emitted at this temperature did not burn when they came into contact with air. This view indicates that if combustible gases are present, their percentage concentration must be small. Measurements were also made of the temperature of the gases escaping from the fumeroles located on the older Zapichu flow; a maximum of 640 °C was recorded.

Guatemala

Note should be made of the extensive field and chemical investigation by E. G. Zies of the domes of the active volcano Santiaguito and its ancient edifice Santa Maria in Guatemala. Unfortunately, failing health prevented him from publishing those studies, mentioned in a series of nine abstracts. The combination of Zies' analytical skills and H. E. Merwin's keen microscopy had generated a detailed picture of the mixing of magmas and the digestion of individual crystals.

Pantelleria

Chayes and Zies were especially interested in the peralkaline lavas of the island of Pantelleria in the Mediterranean Sea between Sicily and Tunisia.[25] They noted particularly the specimens with *ns* (sodium metasilicate) in their

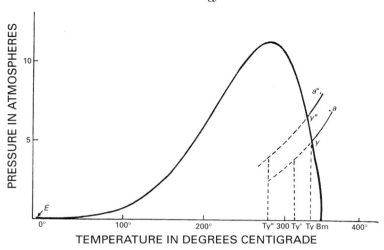

Figure 5.4 Univariant curve for the melting of KNO_3 under various water pressures. This is the origin of the concept whereby crystallizing hydrous magma generates increasing pressure on cooling. *Source*: After Day and Allen, *Volcanic Activity and Hot Springs of Lasen Peak*, p. 78, figure 41B.

norm, those with a high Cl content, and the role of sanidine phenocrysts. Two specimens previously studied by Washington were reanalyzed by Zies.[26] The alkalis were quite similar and removed any of the doubts rumored, facetiously, about their accuracy, presumably arising from Washington's smoking of cigars that were alleged to yield an alkali-rich ash. The molar excess of alkalies over R_2O_3[27] was confirmed and is expressed as *ns* in the norm. It was the opinion of Zies that the sodium metasilicate was in the glass for the most part and did not necessarily depend on the amount of cossyrite present, as Washington had suggested.

Zies also took issue with Washington on the analysis of a basalt from Pantelleria that differed in titania (TiO_2) and alumina (Al_2O_3). The total of TiO_2 and Al_2O_3 were the same in each of their analyses, but Zies used new methods for their difficult separation. Nevertheless, the TiO_2 content of the basalt (3.94 wt.%) was considered high. Zies noted also that there was no adequate method at that time to determine the oxidation state of the titanium. There is little doubt that Washington and his colleagues contributed a substantial amount of new chemical data toward an understanding of the volcanic rocks of Italy.

Explosive volcanism

In the course of the studies on Mount Katmai, Alaska, G. W. Morey provided a new theory for the increase in pressure of a cooling hydrous magma,[28] based on the continuity and univariancy of the crystal + liquid + gas curve in

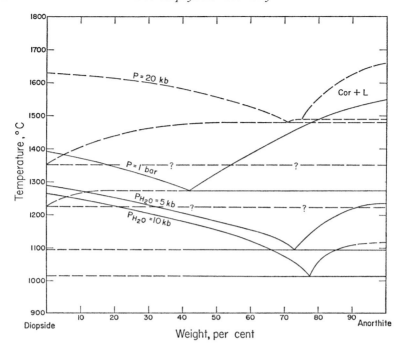

Figure 5.5   Phase relations on the system diopside-anorthite at 1 atm and at P_{H_2O} of 5 and 10 kbar, and an estimate at 20 kbar. The large shift in the isobaric eutectic toward anorthite with increasing water pressure may bear on the origin of anorthosites. *Source:* From H. S. Yoder, Jr., "Diopside-anorthite-water at five and ten kilobars and its bearing on explosive volcanism," *CIW Year Book* 64 (1965), 86, figure 11.

the KNO_3-H_2O system (Figure 5.4).[29] Some forty years later Yoder pointed out that most magmas were not saturated or univariant and that explosive volcanism resulted from an incremental drop in pressure when gas was liberated from an initially undersaturated magma.[30] This concept arose out of an experimental study of the synthetic basalt system diopside-anorthite-H_2O at 5 and 10 kbar (Figure 5.5). A cautious statement by Yoder regarding "the potentially explosive character of the stratovolcanoes of the western United States"[31] was spread over hundreds of newspapers in 1965. The explosive eruption of Mt. Saint Helens in 1980 provided some support for the prediction even though the recommendation was made "to establish observatories near Mount Ranier and Lassen Peak."

6

THERMODYNAMICS

The experimental determination of the stability relations of the minerals in the Earth as outlined by the advisors to the Geophysical Laboratory constitutes an enormous task. The range of conditions of pressure, temperature, and compositions is indeed great, but not beyond present capabilities in the laboratory. There is a technique for augmenting, expanding, extrapolating, and verifying the relations observed through the use of thermodynamic properties first worked out by J. Willard Gibbs in 1876.[1] Unfortunately, his theoretical treatment required that the thermodynamic properties must be determined experimentally in order to calculate the stability relations of the assemblage of minerals. Members of the Geophysical Laboratory set out to evaluate as many as possible of the thermodynamic relations in multicomponent systems and to tabulate them in a form readily available for general use. In addition, it was necessary for them to expand Gibbs' skeletal outline of the theory to achieve formulation of all possible relations between all the variables involved.

The first director, A. L. Day, while attending Yale occupied an office in the same building as J. Willard Gibbs, and presumably Gibbs influenced Day's scientific focus. Later, on returning from Germany, Day served as the personal emissary of the Berlin Physical Society, advising Gibbs of his election as president of the Society, which Gibbs declined for reasons of age. It is no wonder, therefore, that thermodynamics became a major factor in the work of the Geophysical Laboratory. Members of the staff conducted intensive study sessions, and several members achieved international recognition for their interpretation and application of thermodynamic principles. After a series of papers on laws of chemical equilibria, heterogeneous equilibria, and phase rule problems, G. W. Morey was asked to contribute an article to the *Commentary on the Scientific Writings of J. Willard Gibbs* (1936).[2] Another more recent major contributor to thermodynamics was George Tunell, renowned for his careful exposition and explicit derivations, particularly in regard to open systems.[3]

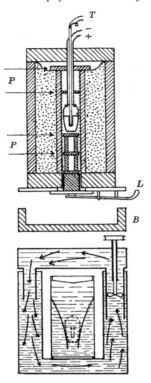

Figure 6.1   W. P. White's calorimeter. Porcelain tube with heating element contains crucible with sample in which thermal element T is imbedded. Box B is removed, latch L releases crucible supports, and current is applied to melt support wire. Crucible drops into open calorimeter, containing water at room temperature, and comes to rest in position indicated by dotted outline. Temperature change in water is a measure of the heat capacity of the heated sample. *Source*: From W. P. White, "Specific heats of silicates and platinum," *Am. J. Sci.* 28 (1909), 337, figure 1

Calorimetry

As the name implies, thermodynamics deals with heat and work. The change of state of one system to another usually involves conversion of heat into work, or vice versa. For this reason, the early work focused on calorimetry wherein the amount of heat can be measured by the consequential change in temperature. In principle, a mineral is heated in a furnace in a platinum crucible that is dropped into a jacketed container of water stirred at room temperature and the rise in temperature measured with multiple thermoelements (Figure 6.1). By dropping the mineral held at a range of temperatures below 1500 °C, the specific heat can be determined. W. P. White carried out the first measurements in 100 °C intervals for diopside, orthoclase, wollastonite, psuedowollastonite, quartz, and two glasses, publishing his results in 1909.[4] After consideration of the various sources of error, he believed the values were better than 0.5% at most temperatures.

In the following twenty years, White devoted his attention to increasing the accuracy of the specific heat measurements. The problems of thermal leakage, thermometer lag, evaporation, insulation, heating methods, and convection were all addressed, and White believed he had attained an "average error of 0.1 per mille".[5]

Another thermodynamic property – heat content or enthalpy – is also critical to the calculation of the stability region of minerals. The heat content is obtained by dissolving the mineral in a solution calorimeter using, for example, hydrofluoric acid (20%). A solution calorimeter (Figure 6.2), designed after one used by Torgeson and Sahama,[6] was constructed by Kracek *et al.* with the intent of resolving the stability of jadeite, as reported in their work of 1951.[7] With stirrers, resistance thermometers, thermostatic oil bath at 75 °C, radiation shielding, and a gold calorimeter vessel, they were able to obtain the heat content for all the minerals presumed to be involved in the reactions to form jadeite. The bottom line was that jadeite formation was thermodynamically feasible at 25 °C without the application of pressure, contrary to geological field interpretations.

It is most difficult to obtain heats of melting because of the reluctance of most silicates to crystallize once melted, particularly in a calorimeter operated near room temperature. With a high-temperature calorimeter H. S. Roberts was able to measure directly the energy actually absorbed while the silicate melted (Figure 6.3).[8] For demonstration he obtained data on pure K_2SO_4 (melting point 1069 °C) and an impure Na_2SiO_3 (melting point 1089 °C). He next tried the important mineral albite in its natural form, but the results differed widely from the calculated heat of melting of synthetic albite obtained by Bowen in 1913[9] from the system albite–anorthite. Subsequently, Roberts measured directly the heat of melting of a 50:50 albite–sodium disilicate mixture and obtained a value closer to Bowen's calculation.

Volume

One of the critical parameters in thermodynamic calculations is the volume of a substance. Of particular importance to geophysics is the change of volume with pressure (compressibility) and with temperature (thermal expansion). The earliest grants from CIW for compressibility studies were made to investigators at other institutions: T. W. Richards and W. N. Stull at Harvard University, who published their work in 1903;[10] G. F. Becker, (1905) at US Geological Survey;[11] and F. D. Adams and E. G. Coker (1906) at McGill University.[12] The materials studied included liquids, gases,

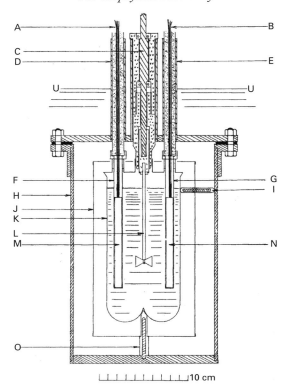

Figure 6.2   Hydrofluoric acid solution calorimeter of Kracek, Neuvonen and
Burley. A, B: leads from the resistance thermometer and calibrating heater; C: stirrer
assembly; D, E: chimneys in the cover of the calorimeter jacket; F, G: protecting
gold wells for the measuring instruments; H: calorimeter jacket; I: centering
supports (three) for the calorimeter and radiation shield; J: radiation shield; K: gold
calorimeter vessel; L: stirrer propeller shaft; M, N: resistance thermometer and
calibrating heater; U: level of immersion in thermostat filled with oil; O: insulating
support for calorimeter and shield. From F. C. Kracek, K. J. Neuvonen, and
G. Burley, "Thermochemistry of mineral substances, I: A thermodynamic study of
the stability of jadeite," *J. Wash. Acad. Sci.* 41 (1951), 375, figure 1. By permission of the
Washington Academy of Sciences.

minerals, rocks, and glass. Their results indicated that compressibility
decreased with pressure, and that granite was more compressible than basalt.
Of special interest is the use of the Washington Monument elevator shaft
by Becker to obtain the elasticity of copper and steel wires, some 480 feet in
length with successive loads with due regard for the time delay.[13]

J. Johnston was concerned with the "flow" of metals by real melting,
and came up with the idea that there were some parallels in their elastic
behavior.[14] He noted that glasses behave as liquids with exceedingly high

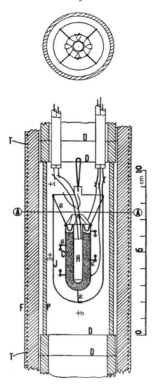

Figure 6.3 High-temperature calorimeter of Roberts. Charge is held in a platinum crucible (C) provided with a central well for the electric heater (H). The calorimeter heater (F) is maintained at a specific temperature below the melting temperature of the charge, and the rise in temperature due to the latent heat of melting is measured outside the jacket (J). From H. S. Roberts, "Direct measurement of silicate heats of melting," *Am. J. Sci.* 35A (1938), 276, figure 2. Reprinted by permission of the *American Journal of Science*.

viscosity. In this way he reasoned that metals flow "not of the whole mass of metal at any one instant but of successive groups of particles." Much more recently, in the 1990s, finite clusters were also used by Cohen and Gong in their development of the physics of melting from first principles.[15] Large-scale numerical simulations of high-pressure melting in MgO revealed that displacements of ions reached about 18% of the near-neighbor distance over a pressure range of 300 GPa. Their results indicate that melting is related to intrinsic instability of crystals and that the oxygen ion may dominate melting behavior at high pressures. In their view, high pressure favors efficient packing of the liquids as well as the solid. Again using molecular dynamic simulations, Cohen and Weitz found that diffusion rapidly increased in MgO to near liquid value below the melting point and described it as "premelting of MgO."[16] Such behavior has implications for lower mantle rheology where diffusion is thought to be the dominant deformation mechanism.[17] Because $(Mg,Fe)SiO_3$ perovskite comprises perhaps two-thirds of the mineralogy of the lower mantle, Marton, Ita and Cohen investigated its equation of state and obtained density and velocity profiles in agreement with seismological models for a reasonable geotherm.[18]

It was also the view of L. H. Adams in the late 1930s that "volume is the oldest, the simplest, and the most fundamental of all the physical attributes."[19] A substantial "bomb" was constructed by Adams, Williamson, and Johnston in 1919 to determine the compressibilities of rocks and their constituent minerals up to 12 kbar, corresponding to a depth of 40 km into the Earth.[20] In 1923 Adams and Williamson concluded that basic or ultrabasic material occurs at a relatively small depth below the surface,[21] and in 1929 Adams and Gibson suggested that the composition of the Earth was sequentially basalt, eclogite, and peridotite.[22] Although their focus was on relating compressibility to the velocity of earthquake propagation at depth, they were also concerned with tidal deformation, critical loading, and the effect of pressure on the stability of minerals.

In succeeding years the compressibility of spinel, analcite, and magnetite, as well as solutions and gases such as H_2S, H_2, was investigated. In the 1970s, from a crystal structure study of fassaitic clinopyroxene, R. M. Hazen and L. W. Finger suggested that the packing of oxygen atoms becomes more regular at high pressure, and they proposed a program to calculate polyhedral volumes from a set of atomic coordinates and a unit cell.[23] On the basis of the variation of cell parameters of wüstite with pressure, B. Siemens and F. Seifert concluded that it would be unlikely for wüstite as a principal phase in the mantle to become stoichiometric under any conditions of pressure and temperature.[24]

The compression of the gases Ar, Ne, and CH_4 led to crystallization at pressures up to 90 kbar, and Hazen and co-workers in 1980 recommended their sharp freezing points as ideal hydrostatic pressure standards.[25] In addition to providing a rigorous derivation of equations for the adiabatic gradient, D. Rumble, a few years earlier, had exhibited the adiabats on a *P–T* diagram for the system albite-H_2O. He contrasted the different behavior of adiabatic decompression along a univariant curve that results in crystallization with adiabatic decompression in a divariant region (L+X) that results in melting. He concluded that divariant decompression is not an impediment to convection, whereas univariant decompression would tend to freeze rising convection currents.[26] An especially critical paper on the compression of hydrogen was published by Hemley and colleagues in 1990.[27] Because of its exceptionally high compressibility, several orders of magnitude higher than most common materials, they provided information on the evolution of properties in general with pressure.

Theory

Major extensions and amplification of the theory of thermodynamics was undertaken by GL staff members in order to make readily available the relations in multicomponent systems. Many of the intervening steps omitted by Gibbs were treated more fully and completely. Instigated by the condensed thermodynamic formula derived by P. W. Bridgman (1925),[28] Goranson prepared an extensive compilation for variable mass functions for multicomponent systems. The evaluation of the mathematical functions was issued by CIW as Publication No. 408.[29] This collection was reissued in 1985 as Publication No. 408B when Tunell addressed the erroneous assumption made by others that led to errors in Goranson's basic equations for energy and enthalpy.[30] Because of Tunell's interest in ore deposits, he also corrected the thermodynamic relations in open systems in a special CIW publication in 1977 (No. 408A).[31] This important contribution on open systems and irreversible processes resulted from his delivery of a set of lectures in the spring of 1975 at the Geophysical Laboratory. Previously, Tunell had prepared a paper on history and analytical expression of the first and second laws of thermodynamics. In 1935, Gibson assembled, but did not comment on critically, the large number of papers published by American chemists on thermodynamics and thermochemistry.[32]

L. H. Adams recognized the utility of the thermodynamic quantity called "chemical potential," but thought "activity" gave a clearer picture of the change in properties of a system.[33] He defined with a single equation the equivalent of the two equations of the usual definition. He provided correlations between activity coefficient, osmotic coefficient and related functions and their derivatives with respect to temperature and pressure. The difference between the two types of line integral in thermodynamics was elucidated by Tunell in 1941.[34] Whereas energy and entropy depend only on the initial and final states, the line integrals for work and heat depend on path. A very useful contribution was made by Goranson to the *Handbook of Physical Constants* (1942) edited by Francis Birch, in which he tabulated the heat capacity and heat of fusion for minerals and rocks.[35]

Applications

Several examples are given here to illustrate how thermodynamics has been useful in providing guidance for new experiments, expanding the relationships, and testing available experimental data.

Because of the many failures to synthesize jadeite ($NaAlSi_2O_6$),[36] an alleged "high-pressure mineral," efforts were made to obtain the requisite thermodynamic data to calculate its stability range. With values of entropy, enthalpy, thermal expansions, compressibility, heats of solution, and specific heats, the

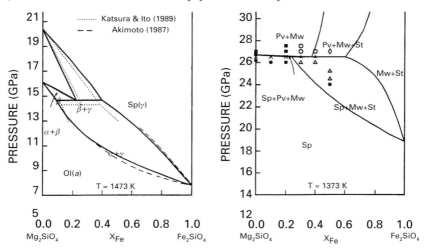

Figure 6.4   Isothermal phase relations at 1873 °K for Mg$_2$SiO$_4$–Fe$_2$SiO$_4$ at two different ranges of pressure. The solid curves are calculated and the dotted and dashed curves represent the experimental data. The data points at the higher pressures are from Ito and Takahashi (see n. 43). Ol, olivine; Sp, spinel; Mw, magnesiowüstite; Pv, perovskite; St, stishovite. From Y. Fei, H-k. Mao, and B. O. Mysen, "Experimental determination of element partitioning and calculation of phase relations in the MgO-FeO-SiO$_2$ system at high pressure and high temperature," *J. Geophys. Res.* 96 (1991), 2167. By permission of the American Geophysical Union.

change of free energy for the reaction nepheline + albite = 2 jadeite was computed mainly from unpublished data by H. S. Yoder and C. E. Weir. They concluded, in 1951, that pressure is not required for the formation of jadeite even at room temperature.[37] Pressure does, however, favor the formation of jadeite at a slowly decreasing rate. Fortunately, a few years later, Coes was successful in synthesizing jadeite at very high pressures.[38] Extrapolation of the experimental *P–T* curve of Robertson *et al.*[39] supported the conclusion that jadeite was indeed stable at or near room temperature and pressure. Additional support for this view may be found in the occurrence of jadeite in a vein associated with serpentine.[40] L. H. Adams held the view that jadeite is found in regions with lower than normal temperature gradient, or has been produced as an unstable phase and persisted at pressures lower than required for its stability.[41]

Critical to an understanding of the minerals in the mantle is the determination of the phase relations in the MgO-FeO-SiO$_2$ system. The partitioning of Mg-Fe between magnesiowüstite, olivine, spinel, and perovskite was experimentally measured in 1991 by Fei *et al.*[42] and a solution model was developed for this partitioning. In turn, they calculated the phase relations (Figure 6.4) and compared the results with those found experimentally by others for specific portions of the curves.[43] Such computations provide critical information on the relations between bulk chemical composition, stable mineral assemblages, geotherm and density profiles through the mantle. They cautioned, however, that to make comparisons with seismic data, at least Ca and Al should be included in the system.

7

X-RAY CRYSTALLOGRAPHY

The program in crystal-structure determination was initiated at the Geophysical Laboratory in 1919,[1] just seven years after Laue's discovery of X-ray diffraction. Prior to 1919 crystal-structure work in the USA had been limited to CIT, MIT, Cornell University, and General Electric Co. The first American crystal-structure determination based on single-crystal data on chalcopyrite was published by Burdick and Ellis in 1917.[2] The work was actually carried out at MIT under Professor A. A. Noyes supported by Carnegie Institute of Washington grants from 1903 to 1927. Noyes held joint appointments at CIT and MIT from 1913 to 1919, but transferred his studies to CIT in 1919 after resigning his post at MIT. While visiting Cornell, Dr. Day was impressed with the X-ray crystallographic technique and invited Ralph W. G. Wyckoff, a graduate student under the guidance of Professor Shoji Nishikawa, to join the staff of the Geophysical Laboratory in 1919. A small building was constructed in 1922 for Wyckoff at the corner of the Laboratory lot on Upton Street to house an X-ray laboratory. Presumably there was some concern about radiation leakage as the room housing the two X-ray generators was completely lead lined (Figure 7.1). The building was used as an X-ray laboratory until about 1955 when it was converted to living quarters for pre- and post-doctoral fellows.

Atomic arrangements were initially deduced intuitively and then tested by a few measured reflections. In the early 1920s Wyckoff derived a complete analytical expression of Schoenflies' space group theory to define all possible arrangements and used the X-ray information to select the correct structure.[3] This work was the forerunner of the "International Tables of X-ray Crystallography" in which the types of atomic positions in the 230 space groups are designated by their Wyckoff letters. Using this method, Wyckoff[4] was able to work out the relatively simple structures of such minerals as dolomite, alabandite, aragonite, barite, periclase, quartz, wüstite, diopside, cristobalite, and zircon as well as many halides in a five-year period. In his experiments Wyckoff utilized three principal techniques: (1) diffraction from single-crystal specimens, (2) diffraction from crystalline powders, and (3) passing the X-rays through a thin section of a crystal. The first structure to be determined at high temperature from powder X-ray diffraction data was high cristobalite.[5] In a study of ammonium chloroplatinate, Wyckoff and

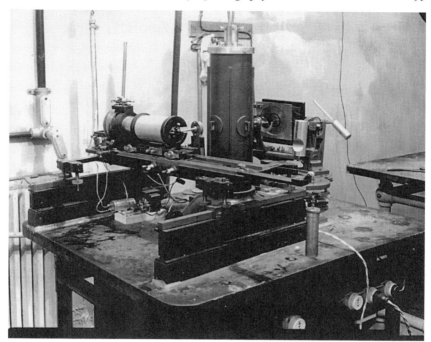

Figure 7.1   Single-crystal, X-ray cameras. Weisenberg on the left and Buerger precession on the right (Archives, GL).

E. Posnjak proved conclusively the validity of Werner's coordination theory.[6] With the help of Wyckoff, Posnjak designed a twin-gas tube for X-ray generation (Figure 7.2). Interspersed with these activities were studies on organic compounds. It is alleged that Wyckoff left the Geophysical Laboratory in 1927 for the Rockefeller Institute because the Director of the former would not support Wyckoff's wish to focus on organic materials.

Before Wyckoff left the Laboratory in 1927, he was joined by an outstanding research associate, S. B. Hendricks, who had been trained under Linus Pauling as his first graduate student at CIT in Noyes' Chemical Laboratory. Although officially associated with the Geophysical Laboratory for only one year, Hendricks continued the collaboration while at the Fixed Nitrogen Laboratory to cooperate with staff members throughout the 1930s. "In 1931 and 1932 Hendricks, Kracek, and Posnjak verified Pauling's hypothesis that in sodium nitrate and ammonium nitrate, molecular rotation takes place in the solid state. This was another new phenomenon first described at the Geophysical Laboratory."[7]

Posnjak was joined in 1929 by T. F. W. Barth on a study of the spinel problem, which resulted in their recognition of the phenomenon of "variate

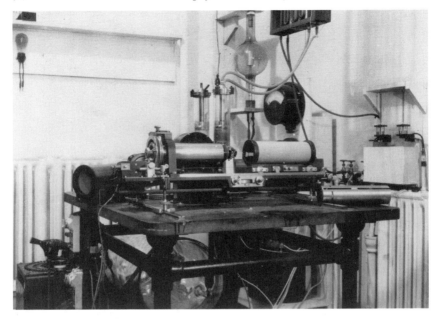

Figure 7.2   Gas-type X-ray tubes with Weisenberg cameras (Archives, GL).

atom equipoints." In 1931 they described how crystallographically equivalent sites could be occupied by chemically different atoms.[8] This demonstration of cation disorder was the key idea in understanding many crystal structures, especially the aluminosilicates, in which Al and Si are often disordered on sites coordinated to four oxygens.

The sulfide minerals were the focus of attention of George Tunell. He determined the structures of tenorite, calaverite, sylvanite, bradleyite, and krennerite.[9] He also derived the Lorenz correction factor for equi-inclination Weissenberg films, essential for the determination of corrected intensities of diffracted X-rays. The Patterson-Tunell stencils and strips (1942) were a very popular aid for the computation of Fourier transforms prior to the computer era.[10]

By 1933 the X-ray laboratory was so well equipped that it attracted Professor J. D. H. Donnay of Johns Hopkins University, and Professor Palache at Harvard sent his assistant H. Berman for instruction on the new equipment. The productive association resulted in the formation of the "Calaverite Club," composed of C. Palache, M. A. Peacock, J. D. H. Donnay, G. Tunell, and T. F. W. Barth. Papers on calaverite, a mineral whose morphology apparently violates the Law of Rationality, were written by members of the club. A joint paper was published in 1934 by J. D. H. Donnay, Tunell, and Barth on "Various modes of attack in crystallographic investigation."[11] Tunell also

Figure 7.3   Norelco powder, X-ray goniometers. First installed in 1949 and in continuous use (Archives, GL).

teamed up with C. J. Ksanda, a chemist, and H. E. Merwin, an optical crystallographer, to characterize an array of compounds until World War II set in.

Attention switched briefly from single-crystal structures to powdered mineral identification with the arrival of the new Norelco powder X-ray diffractometer in 1949 (Figure 7.3). After extensive calibration tests by L. H. Adams, then director of the Geophysical Laboratory, and its automation with a turntable constructed by J. Van den Heurk,[12] it was used round the clock for identification of the products from the numerous phase-equilibria runs made by the very efficient Dr. J. F. Schairer on pure mineral systems.

With the arrival of Gabrielle Donnay,[13] from MIT, in 1950, the first woman to receive a Ph.D. in X-ray crystallography, the attention returned to single-crystals. She was highly productive both with Fellows and cooperatively with her husband, Professor J. D. H. Donnay, after her staff appointment on January 1, 1955. Professor J. D. H. Donnay became a Guest Investigator in 1959, a relationship that continued until 1970 when they both left to join the faculty at McGill University. The Donnays were especially successful in relating morphology to structure, which led to additional generalizations of the Law of Bravais. They found a high-temperature symmetry change in the alkali-feldspar series,[14] defined the magnetic structure of chalcopyrite, pursued the array of solid solutions in tourmaline[15] and nepheline, and predicted the structure of the one-layer micas. In addition to discovery of a new allotropic form of carbon in the Ries crater, G. Donnay with D. L. Pawson found the relationship between the crystallographic axes and morphological features of the calcite skeleton, termed biocrystals, in Echinodermata.[16] The Donnay team also became known internationally for their pioneering efforts with others to compile and systematize the findings of all crystallographers. The first edition of "Crystal Data" (1954) had 719 pages and a second edition (1963) had 1302 pages, all prepared by hand.[17]

In the course of a study of the bornite-digenite join, Cu_5FeS_4-Cu_9S_5, by N. Morimoto and G. Kullerud, the existence of various superstructures made it necessary to examine the crystals at high temperatures.[18] The Buerger precession camera was modified by Morimoto and J. L. England for use to 300 °C.[19] It was thereby possible to confirm the existence of three different forms of bornite and digenite by means of this modified camera as well as to obtain data on their high-temperature polymorphs.[20]

During the Donnay period, two outstanding postdoctoral fellows, J. V. Smith (1951–1954) and C. W. Burnham (1961; 1963–1966) also contributed to the advancement of X-ray crystallography. Smith developed X-ray methods for determining the composition of natural nephelines, the alkali feldspars, and the soda-rich plagioclases. His characterization of the structural types of the micas is still used today.[21] Burnham, a former student of Buerger at MIT, took on the difficult task of refining the crystal structures of both sillimanite and kyanite (1963).[22] The structure of jadeite was refined by Prewitt and Burnham in 1966[23] and a new structural type in the micas, 3-T, was identified by L. Güven and Burnham a year later.[24] In the fall of 1954, J. V. Smith resigned his fellowship to become a demonstrator at his home university, the University of Cambridge, England; Burnham stayed on to become a regular staff member as petrologist at the Geophysical Laboratory in the fall of 1963.

Both theoretical and practical X-ray crystallography were given another major thrust with the arrival of L. W. Finger, a Fellow, from the University of Minnesota in 1967. He, too, was attracted to the Geophysical Laboratory because it had one of the first Supper-Pace automated equi-inclination single-crystal diffractometers (Figure 7.4). He immediately converted the machine so that it could perform step scans with the count recorded at each step.[25] In addition, he replaced the paper-tape punch with a magnetic tape unit. In 1969 he was appointed Crystallographer on the regular staff. Finger was an expert in the design and construction of computer-automated control systems for X-ray diffraction and electron microprobe experimental systems. He refined a wide range of crystal structures among the common rock-forming minerals, contributing many techniques for resolving the more complex mineral structures. He participated in the identification of the

Figure 7.4 Supper-Pace, automated, equi-inclination, single-crystal diffractometer (Archives, GL).

minerals in the lunar samples and described the new mineral armalcolite from the Apollo 11 samples.[26] The armalcolite structure was later refined by Wechsler, Prewitt, and Papike.[27]

Initially, emphasis was given to the crystallographic details of rock-forming silicates in order to deduce geological history from such properties as twinning, transformations, cation ordering, and exsolution. The chain silicates received the greatest attention, perhaps because their complexity results in the capacity to record more information about their conditions of formation. Investigation of the complex pyroxene and amphibole groups required new procedures for the collection and analysis of X-ray intensity data. Major advances were contributed in both automated data collection and structure refinement.[28] Procedures for the refinement of partial occupancies, cation ordering, and complex thermal vibration were incorporated into the program RFINE, now in widespread use for X-ray and neutron diffraction studies. In his refinement of the structure of anthophyllite, Finger related the occupancies of the four octahedral sites to the temperature of formation, or annealing, in a quantitative way using those new techniques.[29] Clinopyroxenes from a diamond-bearing pipe were annealed by McCallister, Finger, and Ohashi,[30] and a cation ordering versus temperature calibration curve was determined. The clinopyroxene crystals themselves indicated an intracrystalline closure temperature of 530 °C below which the rate of cation ordering was negligible, whereas the temperature below which the intercrystalline partitioning of Ca and Mg between clinopyroxene and orthopyroxene was negligible was 1375 °C.[31] The difference between these closure temperatures, together with the occurrence of nongraphitized diamonds, places important constraints on the rate of cooling of the rock. Other refinements of the common rock-forming minerals (e.g. olivine, pyroxenes, amphiboles) led to the development of time–temperature-transformation plots of special application to other petrological problems.[32] Of special interest was the unraveling by Ohashi and Finger in 1976 of the structural relations of the $(Mn, Mg)SiO_3$ pyroxenoids, which occur in several complex but related chain structures.[33] The differences between bustamite and wollastonite, for example, could be related to the stepwise cation ordering in the octahedral strip.

By the mid-1970s the interests of the crystallographic investigators at the Geophysical Laboratory had shifted to the study of crystals at elevated temperatures and pressures. The structure of sanidine was refined up to 800 °C, and that of diopside up to 700 °C.[34] Both studies illustrated the principle that relatively simple external changes in unit-cell dimensions with temperature change are the result of complex internal structural changes in bond lengths and angles.

Another highly productive and imaginative team was formed when R. M. Hazen, an experimental mineralogist, arrived in 1976. Hazen and Finger initiated single-crystal, diamond-cell techniques[35] that resulted in the accurate measurement of both compressibility (to 200 kbar) and thermal expansion (to 1000 °C). They modified the design of Merrill and Bassett[36] and the additional improvements in the cell greatly increased the quality of the results. In the period 1977–1979, over 70 crystal structures were determined at high pressures jointly by Hazen and Finger. The crystals included the common rock-forming minerals: olivines, pyroxenes, garnets, spinels, layer minerals, zircon, and corundum- and rutile-type oxides. High-pressure phase transitions were also studied in analcite and manganese difluoride.

Single-crystal studies at high pressures were limited to about 100 kbar for lack of a suitable hydrostatic medium. Fortunately, Finger and his colleagues discovered that rare gases, including argon and neon, remain hydrostatic to much higher pressures.[37] Furthermore, these gases crystallize in the diamond-anvil, high-pressure cell to single crystals that are among the most compressible solid substances known. As a result, a single crystal of FeO was pressurized, for example, to 200 kbar in a neon-filled cell. It is possible, therefore, to study the structures of other single crystals immersed in such gases to pressures well above 100 kbar. From these techniques the concept of cation polyhedral analysis as a function of pressure and temperature evolved. The variation of crystal structure with temperature, pressure, and composition was treated comprehensively in a book by Hazen and Finger published in 1982.[38]

The new opportunities in synchrotron X-ray crystallography were vigorously investigated with the arrival of C. T. Prewitt, an X-ray crystallographer, as Director of the Geophysical Laboratory in 1986. The adaptation of synchrotron radiation to structure determinations of very small crystals held at pressure in the diamond cell resulted in the characterization of phases with high-temperature, superconducting properties. Much excitement was generated in 1988 when Mao and co-workers obtained the crystal structure and equation of state of solid hydrogen to 265 kbar at room temperature using new synchrotron X-ray diffraction techniques.[39] Later, in 2000, Strzhemechny and Hemley compressed hydrogen to 580 kbars and developed a new ortho-para conversion mechanism in the dense solid.[40] The increasing anisotropy provided insight into the molecular-to-atomic transition predicted at even higher pressures. Even though the metallic state, predicted by others, has not yet been achieved, the changes may yield other information of greater theoretical value.[41] These studies and the use of a large array of spectral tools of modern day physics and chemistry led to the focus in the 1980s on the field of mineral physics, described in detail in Chapter 15.

The hallmark of the X-ray crystallographers at the Geophysical Laboratory has been their cooperative response to the needs of other staff members in characterizing both natural and synthetic phases they discovered. A particular challenge has been the small size of the crystals, whether from a meteorite or synthetic product. The detailed variations in even the common rock-forming minerals have provided information of immense value in defining their history of formation. There is no doubt that X-ray crystallography has contributed greatly to the success of the Geophysical Laboratory.

8

SILICATE LIQUID STRUCTURE

The principal agents of mass and heat transfer in the Earth are magmatic liquids. Volcanic and igneous rocks involve a precursor magma in part or entirely liquid. Attention was devoted in the beginning to the origin of this liquid and its crystallization products as a function of bulk composition, temperature, and pressure. By 1950 it was realized that knowledge of the chemical, physical, and thermodynamic properties of silicate melts were critical to an appreciation of their behavior. The liquidus phases, viscosity, element partitioning, immiscibility, electrical conductivity, seismic response, thermal expansivity, density, redox, equilibria, volatile solubilities, diffusion, and compressibility are all dependent on the structure of the liquid.

Initially, data were collected on glass, thought to be the structural equivalent of the liquid. Many of the principles that govern glass and silicate liquid structures were established in 1980 and dealt with its anionic constitution.[1] Other studies demonstrated the close structural relationships between melts and glasses with high-temperature Raman spectroscopic investigations.[2] No coordination changes with temperature and no new anionic species were observed in the melt relative to those in glass. There were, however, changes in the relative proportions of the structural units with temperature, and the glass transition was recognized as a kinetic barrier.[3] The time-dependent response of the structural units and properties in the liquid and glass were termed relaxation. The time scale of this relaxation can thereby be used to reconstruct the temperature dependence of the equilibria in the liquids. With the determination of the temperature of the glass transition and the observed thermodynamic changes associated with the glass transition, the need for studying liquid properties under molten conditions became necessary. The shift in boundary curves of silicate systems as components were added had already been attributed by I. Kushiro to structural changes in the melt.[4] All of the important variables affecting liquid structure have eventually been investigated by staff members of the Geophysical Laboratory. In accord with the general philosophy of the Laboratory, the simplest system was studied first and additional components added one at a time to identify their effects on the structure.

The structure of vitreous and molten SiO_2 was first thought, in the 1930s, to be a random network of SiO_4^{4-} tetrahedra.[5] Thirty years later, a pseudocrystalline structure was proposed,[6] but eventually, in 1982, it was determined that the spectroscopic data best fit a model of coexisting discrete structures, such as interconnected three-dimensional rings with four and six tetrahedra.[7] Addition of metal oxides to silica results in the formation of nonbridging oxygens whose proportions are relevant to all the properties of silicate melts. The specific effects of the principal rock-forming elements on the structure of silicate liquids have been investigated.

Alkalis

As a result of a major breakthrough in technique in the early 1990s, B. O. Mysen and J. D. Frantz were able to obtain high-quality Raman spectra of alkali-metal–silica systems at magmatic temperatures.[8] The structural units found resembled SiO_3^{2-}, $Si_2O_5^{2-}$ and SiO_2. The effect of temperature on the abundance of the structural units was observed and the number of $Si_2O_5^{2-}$ units decreased with increasing temperature. In general, this effect decreased in the order K > Na > Li, and quantitative values have been determined.[9]

Aluminum

The three-dimensional network structures of glass in the systems $NaAlO_2$-SiO_2, $CaAl_2O_4$-SiO_2, and $MgAl_2O_4$-SiO_2 were published by Seifert, Mysen and D. Virgo in 1982.[10] They found mixtures of three-dimensionally inter-connected rings in the alkaline earth system with (1) six-membered SiO_2 rings with or without Al^{3+}; (2) four-membered rings with Al/Si = 1; and (3) six-membered AlO_2^- rings with no Si^{4+}. The proportions of the rings depended on the Al/(Al+Si) ratio. In the $NaAlO_2$-SiO_2 system the struc-ture of the SiO_2 framework is largely preserved and the Al^{3+} is incorporated mainly in the six-fold rings of SiO_2. On the basis of these data, they estimated the relative abundance of the three-dimensional structural units in natural magma. The more basic magmatic liquid has a larger proportion of Al-free SiO_2-rings relative to other three-dimensional network units in the melt.[11]

Titanium

Laser Raman spectroscopy was also used to study the influence of TiO_2 on the structure of silicate melts. The proportion of Ti-rich structural units increases relative to the Si-rich units with increasing TiO content of the system. Mysen, F. Ryerson, and Virgo used melt compositions of Na_2SiO_3, $CaSiO_3$, $CaAl_2Si_2O_8$, Na_2TiO_3, $CaMg Si_2O_6$ and $NaAlSi_3O_8$ to ascertain the role of Ti.[12] All of the titanium was found to be in four-fold coordination. With increas-ing TiO_2 content there is an increase of chain and sheet units relative to monomers. Because of the difference of Ti-O and analogous Si-O bonds it was possible to determine from the spectra the bulk monomer/chain/sheet ratios.[13]

Phosphorous

In all the melts studied by Mysen, Ryerson, and Virgo, the P^{5+} was in tetrahedral coordination. Of all the oxides added to the melts, P_2O_5 had the most profound effect on the liquidus boundaries.[14] Trace element partitioning between immiscible acidic and basic melts in equilibrium was strongly influenced by P_2O_5 according to Watson.[15] For example, REE phosphates in basic melt apparently are more stable from those in acidic melts.[16] Phosphorous is dissolved in quenched SiO_2 melt as discrete sheets of P_2O_5 composition. In three-dimensional aluminosilicate melts (e.g. albite, anorthite), the $AlPO_4^{\circ}$ complexes are formed together with sheet-like P_2O_5 complexes.[17] The resulting structure consists of one P_2O_5 sheet, two (Si, Al) three-dimensional aluminum phosphate units, and a unit with Si and Al in tetrahedral coordination.

Iron

With [57]Fe Mössbauer absorption spectroscopy, Mysen and Virgo[18] studied the melt structures in the system $NaAlSi_2O_6$-$NaFe^{3+}Si_2O_6$. Because Fe^{3+} may not substitute for Al^{3+} or Si^{4+} it forms isolated tetrahedra, and because Fe^{2+} may resemble the network modifiers such as Ca^{2+} and Mg^{2+}, it was essential that the oxidation state be determined as a function of melt composition, temperature, and pressure.

Raman spectroscopic measurements had been carried out on quenched melts on the join Na_2SiO_3-$NaFe^{3+}Si_2O_6$ where the Mössbauer data indicated that within the detection limit of the Mössbauer technique all iron was present in the ferric state.[19] At magmatic temperatures no in situ spectra could be taken on samples with ferrous iron because of oxidation of Fe^{2+} to Fe^{3+} due to surface heating by the laser, photo-oxidation, or both. Consequently, Raman data for iron-bearing alkaline-earth metasilicate melts could not be acquired. In the system Na_2O-Al_2O_3-SiO_2-Fe-O, as well as others, Mysen and Virgo found the Fe^{2+}/Fe^{3+} ratio decreases linearly with increasing $Al/(Al+Si)$.[20] The inverse correlation was more pronounced at higher temperatures. The degree of polymerization of the melts, therefore, was greatly dependent on the oxidized iron relative to the total iron. They showed how the oxidation influenced the viscosity: the viscosity decreased with increasing temperature commensurate with iron reduction. Other properties in addition to viscosity also depend on the degree of polymerization, such as crystal-liquid element partitioning[21] and liquidus phase relations.[22]

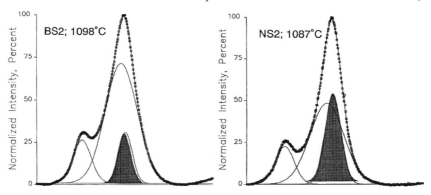

Figure 8.1 Comparison of the relative intensities of the shaded bands representing the $Si_2O_5^{2-}$ structural unit in barium di silicate (BS2) at 1098 °C, and in sodium disilicate (NS2) at 1087 °C. From J. D. Frantz and B. O. Mysen, "Raman spectra and structure of BaO-SiO$_2$, SrO-SiO$_2$, and CaO-SiO$_2$ melts to 1600 °C," *Chem. Geol.* 121 (1995), 175, figure 10. Reprinted with permission of Elsevier.

Alkaline earths

A high-temperature Raman spectroscopic study of the BaO-SiO$_2$ system by Mysen and Frantz revealed that no new structural units formed as the transformation of glass to melt took place.[23] The proportion of the units changed as in the case of the alkalis where the $Si_2O_5^{2-}$ unit decreased with increasing temperature. Later studies included SrO-SiO$_2$ and CaO-SiO$_2$.[24] Little difference was found in the structural behavior; however, the relative abundance of $Si_2O_5^{2-}$ units appear to be less in the alkaline-earth silicates than in those containing alkalis (Figure 8.1).

Volatiles

Vibrational spectra of quenched hydrous aluminosilicate melts indicate that nonbridging oxygens are formed as H$_2$O is dissolved and the OH groups are associated with Si^{4+}, but probably not with Al^{3+}.[25] There was no evidence for metal-hydrogen and H-H bonds. The solubility of CO$_2$ is about one half to one order of magnitude less than H$_2$O. It dissolves as discrete CO$_2$ molecules and as carbonate. The extra oxygen is obtained by transferring two nonbridging oxygens to a bridging oxygen and CO$_3^{2-}$. Mysen and Virgo found evidence that an amount of Al^{3+} in the three-dimensional network equivalent to the proportion of metal cation used to form CO$_3^{2-}$ became transformed from a network former to a network modifier.[26] Fluorine dissolved to form F$^-$ ions that are electrically neutralized with either Na$^+$ or Al^{3+}.[27] There was no evidence for association of F with Si^{4+} or Al^{3+} in four-fold coordination nor in six-fold coordination with Si^{4+}. On the basis of the relative intensities of the Raman bands, they were able to calculate the relative abundance of the structural units.

The solubility of SO_2 is lower than CO_2 and probably reflects the larger size of the SO_2 molecule. Sulfur solubilities in silicate–sulfide melts are highly dependent on the oxygen fugacity, decreasing with increasing oxygen fugacity at high pressures.[28] Sulfur dioxide probably enters silicate melts through the formation of SO_3^{2-} and SO_4^{2-} complexes. In albite melts the solution of SO_2 produces Al^{3+}, Na_2SO_3, SiO_2, and $Si_2O_5^{2-}$.[29] Other depolymerized units may be substituted for the disilicate unit, but more study is required. A small amount of data on using ^{35}S and on the solubility of SO_2 in albite melt[30] indicate behavior similar to CO_2.

As outlined above, the composition of a melt greatly influences the kinds and proportions of the structural elements. In turn, the liquidus phase relations, thermochemical properties, viscosity, volume properties, and almost all other liquid properties are determined by the structural units. Since 1992 it has been possible to conduct liquid structure experiments at magmatic temperatures, and the quantitative aspects are now one of the major goals of liquid-structure research. Complimentary to the experimental determination of the structural units and properties, programs are underway to calculate the degree of polymerization and strength of bridging oxygen bonds that characterize the physical properties. The thrust of this high-temperature work has been the examination of the relationships between electronic properties of metal cations (alkalies and alkaline earths) and the speciation in the melt as well as the relationship between melt structure and form of charge balance of Al^{3+} in tetrahedral coordination. The purpose is to rationalize melt structure and properties governed by the configurational states of melts, properties governed by both Al-Si and metal cation order–disorder relations, and facets of liquidus phase relations and element partitioning.

9

ORE GEOCHEMISTRY

The plans submitted in 1903 by Dr. C. R. Van Hise for a geophysical laboratory included the need for experimental studies on the theories of ore deposition. There was concern already at the beginning of the century for the mineral wealth of the nation and the general question of the conditions of formation of sulfide ore bodies in particular. It was realized that such studies would be of "great practical importance in the exploration and exploitation of ores."[1]

The first problem to be met was the accurate determination of sulfur not only in soluble sulfates but also in the sulfides. In addition, sulfuric acid was of great importance in manufacturing. E. T. Allen and J. Johnston provided new techniques for the analysis of sulfates and sulfides and discovered that the sulfides of iron, in particular, oxidized when simply ground in the atmosphere leading to large errors if not attended.[2] Geological observations had already led to the widely held conclusion that vein-forming solutions had precipitated the ores. The relative thermal stabilities of pyrite-marcasite and pyrite-pyrrhotite[3] provided information on temperature of formation. This concept was also applied to the sulfides of zinc, sphalerite and wurtzite.[4] These studies served to clear up the long-mooted question of the true composition of these sulfides and their interrelationships.

The secondary enrichment of copper sulfides was unraveled in 1915 by E. Posnjak and co-workers from a thermal and microscopic viewpoint, and a year later E. Zies and colleagues provided the chemical aspects.[5] The solid solutions of both the natural and synthetic products received attention, particularly in ascertaining the dissociation or melting temperatures. The dissociation pressures of sulfides were studied by Allen and R. H. Lombard who were forced to devise a new method when spiral quartz gauges were not available from Europe because of World War I.[6] Their apparatus (Figure 9.1) essentially compared the known pressure of pure sulfur with that of the sulfide. An example is given for pyrite (Figure 9.2). The studies of Allen and Zies in 1923[7] on the chemistry of hot springs was applied by C. N. Fenner ten years later to ores derived from igneous origins.[8] In the same year (1933), N. L. Bowen proposed that the heavy metals would be concentrated in the residual fractions of a differentiating magma.[9] It became evident that hot

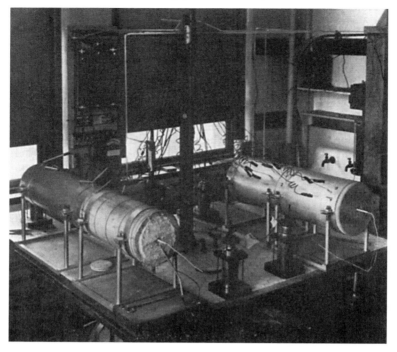

Figure 9.1 Apparatus for determination of the dissociation pressures of sulfides by comparison with the known vapor pressure of liquid sulfur. From E. T. Allen and R. H. Lombard, "A method for the determination of dissociation pressures of sulfides, and its application to covelite (CuS) and pyrite (FeS$_2$)," *Am. J. Sci.* 43 (1917), 180, figure 3.

springs were an end product of ore deposition as well as indicators of volcanic activity.

In accord with the philosophy of investigating silicates in a component by component approach, the ore minerals were initially studied by examining, for example, Fe-O, Fe-S, Cu-O, and Cu-S. A benchmark paper on the Cu-Fe-S system (Figure 9.3) by H. E. Merwin and R. H. Lombard appeared in 1937.[10] They laid out the technique for holding synthetic and natural samples at a defined vapor pressure of sulfur (455 mm) and temperature in silica-glass (vitreosil) tubes. The phase diagram was of great importance to economic geologists because it helped to constrain the temperatures and pressures of ore formation. With continued investigation, the system has been found to be exceptionally complex, and it remains one of the most intensively studied systems even today. The concept of buffers was clearly defined even though the "equilibrium" obtained was of a restrictive type, the buffer and sample being at different temperatures. Exsolution in the Cu-Fe-S system was studied by R. Brett in the 1960s in controlled cooling experiments.[11] Exsolution

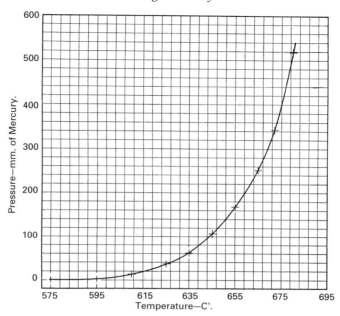

Figure 9.2 Dissociation pressure of pyrite, FeS_2, as a function of temperature (Note 1 atm = 760 mm Hg). *Source*: as Figure 9.1, p. 192, figure 6.

lamellae persisted for as long as seven months, and are, therefore, not exclusively indicative of rapid cooling. In short, quantitative data on the cooling rate or cooling conditions of natural sulfides in the Cu-Fe-S system cannot be obtained through the study of exsolution textures.

The identification of the opaque ore minerals required not only the use of crystallography and chemical composition but also the techniques of the metallurgists, especially for fine-grained assemblages. Polished sections were usually prepared for examination in reflected light.[12] Etching, staining, and fluorescence were occasionally diagnostic. X-ray and electron diffraction and a host of other techniques used by mineral physicists were required for definitive characterization. Identification was also complicated by the difficulty of quenching run products, a problem recognized by Merwin and Lombard in 1937.[13] Conversely, the unmixing of solid solutions was extremely useful in estimating the temperature of formation of natural ores. The rapid response of some of the sulfides to changes in temperature in the laboratory have brought special insight into the metamorphism of ore bodies, a concern of Van Hise as early as 1900. Re-equilibration to temperatures as low as 200 °C for some sulfides has been useful in generating a scale of closure for determining the kinetics of the cooling of the ore body.[14]

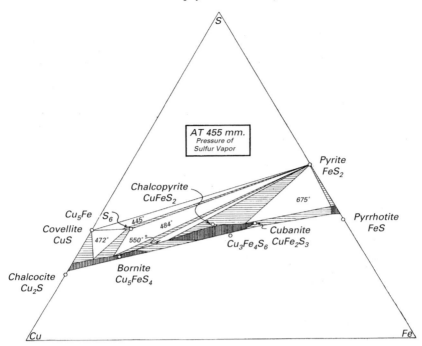

Figure 9.3   The Cu-Fe-S system at 455 mm of sulfur pressure. Vertical ruling represents the extent of liquid compositions. The temperatures mark the three-phase assemblages. The horizontal patterns indicate two-phase assemblages. From H. E. Merwin and R. H. Lombard, "The system, Cu-Fe-S," *Econ. Geol.* 32 (1937), 206, figure 1. Reprinted with permission of *Economic Geology*.

In 1953 G. Kullerud demonstrated quantitatively the unmixing of solid solutions in FeS-ZnS as a function of temperature,[15] thereby establishing the principle of geothermometry. Fortunately, Kullerud came to the Geophysical Laboratory in 1954 and provided major impetus to the study of ore deposits. During a highly inventive period, he evolved techniques for dealing with sulfur and selenium systems, beyond the limitations of silica-glass tubes, up to 1400 °C.[16] Kullerud and his colleagues showed that essentially all sulfide and selenide systems exhibited liquid immiscibility. The first high-pressure-temperature diagram (Figure 9.4) for a sulfide, pyrite, was achieved by Kullerud and Yoder in 1959.[17] Pure gold tubing was used to contain the reactants. Gold does not react with sulfur above approximately 240 °C. The tubes collapse when a specific external pressure is applied. Because there is no free space, vapor is absent and the pressure on the sample is essentially equal to the applied external pressure. Their observation of incongruent melting in pyrite clearly showed that it could not be a magmatic phase in either basalts or rhyolites. A similar conclusion was reached for pentlandite in a study

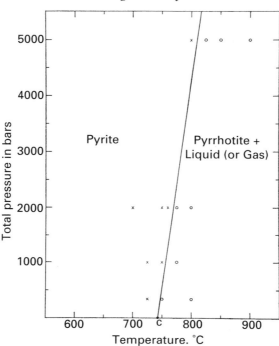

Figure 9.4 Upper stability curve of pyrite, $FeS_2 \Leftrightarrow Fe_{1-x}S + L$. From G. Kullerud and H. S. Yoder, Jr., "Pyrite stability relations in the Fe-S system," *Econ. Geol.* 54 (1959), 554, figure 5. Reprinted with permission of *Economic Geology*.

by Kullerud in 1963.[18] The phase diagram for the economically important system Cu-Fe-Ni-S was established by J. R. Craig and Kullerud in 1969.[19]

The transport of ores in the vapor phase was considered by some to be improbable because of the low temperature of deposition. A study by G. W. Morey and J. M. Hesselgesser in 1951 on the solubility of sulfates, oxides and sulfides in superheated steam in the range of 400–600 °C and pressures up to 2 kbar revealed that the sulfides in particular had low solubility.[20] On the other hand, H. L. Barnes showed that sulfide minerals in an aqueous solution containing reduced sulfur could be transported.[21] He measured the solubilities in ZnS-H_2S-H_2O and concluded sulfide transport in a polysulfide liquid was a potential mode. This concept was correlated by Barnes and Kullerud in 1961 with determinations in the anhydrous Fe-S-O systems.[22] A portion of the Fe-S-O system was studied by A. J. Naldrett in which a ternary eutectic occurs at 915±2 °C.[23] The magmatic sulfide-oxide ores may bear little relationship, however, to such a eutectic point because of the influence of oxygen and sulfur fugacities imposed by the host silicate magma. Those factors have in turn a great influence on the texture of the ore body. Naldrett's work demonstrated how ore magmas can form whether as spherules in some rocks or as interstitial filling in others. On the other hand, the studies of wall rock alteration,[24] fluid inclusions, and phase relations clearly support the formation of hydrothermal ore deposits.

The interrelationships of sulfides and silicates, the essence of ore petrology, did not get underway until the early 1960s. In 1963, Kullerud and Yoder, after considering the zones around the Badenmais ore body (Bavaria) reacted sulfur with fayalite and obtained pyrrhotite, magnetite, and quartz.[25] From these and other experiments with various iron-bearing silicates, the concept of sulfurization emerged that helped explain the apparent high-grade metamorphic aureoles around low-temperature ore bodies. Sulfurization was defined by Kullerud as "The reaction between sulfur from an external source and cations such as iron, nickel, and copper in solid solution in common rock-forming minerals as in igneous magma."[26] This concept was also demonstrated and applied to the interrelationship of sulfides and carbonates as well as sulfides and oxides.[27] In the latter case, Kullerud and colleagues produced a new type of solid solution in magnetite when reacted with sulfur.

The magmatic ores, especially those occurring in the layered igneous intrusions, have been of more recent concern. With an exceptionally wide range of experience in the layered intrusions of the world, T. N. Irvine and his colleagues D. W. Keith and S. G. Todd characterized the Pt-Pd ores of the Stillwater Complex of Montana.[28] The detailed analyses of the J-M reef illustrated how magma mixing and double diffusive convection play a major role in the deposition of ores high in the layered sections of silicate rocks. The commercial drilling of the Skaergaard Intrusion, Greenland, for gold and platinum has intensified Irvine's analysis of that layered intrusion. The focus has been on the upper layers, and Irvine and his colleagues have now developed a theory for their occurrence high in section.[29]

A study of the kinetics of polymorphic transitions in the mercury sulfides was published by N. Z. Boctor and R. H. McCallister in 1979.[30] Very small amounts of Zn and Fe in solid solution retarded the metacinnabar–cinnabar transition in the range of 100–200 °C where most mercury ores form. They described the observations as similar to those predicted by the "impurity drag effect" that influences nucleation and growth mechanisms. The study clearly raised an alarm regarding the use of exsolutions and polymorphic transitions as temperature indicators if impurities are present.

While testing the various models of formation of the Earth's inner core, Li *et al.* found that a significant amount of sulfur appeared to be incorporated in iron at very high pressures, 2.5 Mbar.[31] On the other hand, intergranular films of FeS may account for the sulfur content, and the lattice parameters are too small to correlate with high sulfur content in the metal. The presence of sulfur may account for the lower density of the core and perhaps provide an explanation of the elastic anistropy and shear modulus of the inner core. The constrains of other elements in iron are being actively investigated.

In spite of the fact that some 25 binary systems, 27 ternary systems, ten quaternary systems, and several more complex systems involving ore minerals have been investigated, there is insufficient information to aid in the exploration and processing of ores. The world, especially the USA, the chief consumer of ore minerals, will begin facing shortages within the next 25 years.[32] There is a clear need to establish a national laboratory for basic research in the principles of element concentration with attention to open systems, fluid dynamics, kinetics, thermodynamics, and element partitioning.[33] Ore deposits are not infinite and are not renewable!

10

FIELD STUDIES

Introduction

The important problems in geology are first identified in the field and then experiments are designed to resolve those problems in the laboratory. As Dr. L. H. Adams said in 1937, "Progress in geophysical investigations requires a judicious combination of field work and laboratory experimentation."[1] The historical aspects of geology that provide the effects of past events, however, yield sequential evidence of process not always reproducible in the laboratory. It is probable that Geophysical Laboratory staff members have been in the field in every corner of the globe. For this reason only some of the major studies are cited here to illustrate the concern and need for guidance from field observations.

Volcanoes

The active volcanic regions of Italy were studied by H. S. Washington beginning in 1902 and resulted in a major CIW publication: "The Roman Comagmatic Region" (see Chapter 5). Vesuvius was observed for 18 years by C. S. Perret who attributed the explosive eruption of 1906 to the gas content of the magma. He, too, produced a major CIW publication in 1924 (see Chapter 5). The region was visited by Director A. L. Day in the company of Washington in 1914 who recorded many details of the fumeroles in the crater. No doubt every staff member who has been in the region has visited Pompeii that was buried beneath 6 meters of volcanic ash from Vesuvius. The nuées ardentes and airfall ash buried the town in one day. Excavation of the town has since revealed many features of the gigantic ash deposit and provided insights into the mechanics of its deposition.

The disastrous explosive eruption of Mount Pelée, Martinique, on 8 May 1902 was also studied in the field by Perret. Just six months later, the volcano Santa Maria in Guatemala also exploded and a new edifice called Santiaguito developed. The activity of that new volcano was studied in 1932 and 1935 by staff members. In 1939 the Geophysical Laboratory and the Department of Terrestrial Magnetism (DTM) cooperated in an investigation of the electrical, magmatic, and other properties of the volcano. In 1939, Perret published his observations on the volcano-seismic crisis of 1933–1937 at Montserrat, West Indies, after twelve visits.[2] Continuous records of sulfide gases were obtained at Montserrat and the new shock-recorder was installed at Dominica.

One of the greatest volcanic eruptions in historic times was at Mount Katmai, Alaska peninsula, in June 1912. The material ejected had a total volume of about five cubic miles. At the town of Kodiak, a hundred miles away, the depth of ash was nearly a foot. C. N. Fenner and colleagues focused on the Valley of Ten Thousand Smokes (Figure 10.1), northwest of the Katmai crater, measuring the temperatures and composition of the gases emanating from the fumeroles, collecting the incrustations, and examining the geological aspects of its formation. They contested R. F. Griggs' idea that the Valley was filled with a mud floor,[3] and preferred the idea of an underlying magma fracturing the valley flow and emitting ash from many vents formed on cracks.[4] Fenner called attention to the banded pumice of rhyolite and andesite at Novarupta, which he had assumed was formed from xenoliths.[5] On the other hand, that type of field evidence duplicated at many localities was later used by Yoder in 1973 to support the probability of contemporaneous eruption of highly contrasting magmas.[6] The gases, containing HCl, HF, and H_2S, transported many metals. The reactions producing magnetite encrustations were of concern to E. G. Zies in the late 1920s.[7] One sample was found to be exceptionally pure, and is used to this day for starting material in other experiments. At least ten papers were prepared by Laboratory members on the Expedition to Katmai.

Fumeroles were also studied by Zies at Santiaguito, Guatemala, and at Parícutin, Mexico in the 1940s.[8] At the former, he observed molten sulfur two inches below the surface, and at the latter a blue haze was emitted at temperatures ranging from 350 °C to 640 °C.

Another volcanic region investigated by many members of the Laboratory staff is that of Hawaii. Mention has already been made of the temperature measurements of the lava in Kilauea by Shepherd and Perret in 1911. Day and Shepherd collected gases in the Halemaumau crater in 1911 and 1912. Sosman also obtained gas samples in 1919, and in 1922 Allen reported on his studies of the gases at Sulphur Banks, Hawaii.[9]

During the Pan-Pacific Scientific Conference of 1920, H. S. Washington collected specimens on Oahu and Hawaii and received a large representative collection of rocks from the region from friends. Five papers on the petrology of the Hawaiian Islands were generated from 1923 to 1926 in which many new chemical analyses were given.[10] The same rocks were studied by T. F. W. Barth and detailed microscopic descriptions of the minerals presented in 1931. Particularly helpful was his index of the many uncommon mineral and rock names.[11]

Because of the long historical record of eruptions and exposure of the eruptive sequences, Yoder and C. E. Tilley made an extensive collection of

Figure 10.1 Valley of Ten Thousand Smokes, Katmai, Alaska. Photograph by R. F. Griggs, 1917. *Source:* From R. F. Griggs, *The Valley of Ten Thousand Smokes* (National Geographic Society, 1922), p. 188.

basalts to use in their experimental study of the origins of basaltic magmas, published in 1962.[12] Their field work on the Hawaiian volcanoes was greatly aided by the guidance of Professor G. A. Macdonald who had a large fund of knowledge of the specific outcrops. The staff of the U.S. Geological Survey stationed at the Volcano Observatory on Kilauea has been of great help in assisting the flow of geologists to Hawaii.

Layered intrusions

One of the most intensely studied igneous bodies in the world is the Skaergaard Intrusion, East Greenland. As a prime example of igneous differentiation, the detailed information provided by L. R. Wager and W. A. Deer (1939) gives the sequence of events that produced the layering.[13] The crystallization sequence, crystal accumulation, gravity stratifications, rhythmic layering, convection currents, chilled marginal border rocks. dike swarms, iron enrichment, and the extreme products of differentiation have all attracted debate. For example, the fractionation trend has been questioned by A. R. McBirney,[14] who also established that the chilled margin is not representative of the original magma composition.

Because of the remoteness of the area and the physical challenges, it was necessary to organize expeditions to study the area. Staff member Neil Irvine has formed or participated in at least a dozen such expeditions to make detailed studies of the critical aspects of the intrusion. His previous experiences with the Duke Island Ultramafic Complex in southeastern Alaska,[15] the Muskox intrusion of the Canadian Northwest Territories,[16] and the Stillwater Igneous Complex in the Beartooth Mountains in Montana[17] led to a focus on the critical questions. The mechanisms whereby graded cumulate layers are sorted and deposited by crystal-liquid suspension currents,[18] the rhythmic alternation of uniform layers, the origins of the many autoliths and xenoliths, the metasomatic effects due to the infiltration of water-enriched intercumulus liquid and associated pegmatization, trough generation and trough layering structures, and the heat transfer problems were just a few of the issues addressed.[19] The marginal border rocks were studied specifically by J. H. Hoover in the late 1980s,[20] and he found that the sequence of rocks corresponded to those of the layered series. In short, there was no evidence of extensive fractional crystallization preceding the formation of the exposed layered series. The upper border group of the Skaergaard Intrusion was examined in detail by H. R. Naslund on the expeditions of 1971 and 1974.[21] He called attention to the significant lateral variations in the rocks formed during the late stages of differentiation of the intrusion. He attributed this to lateral inhomogeneities perhaps resulting from the restriction of convection into isolated cells in the late stages.

The largest known layered intrusion is the Precambrian Bushveld Complex in South Africa. It has been studied extensively mainly because of its great economic importance for platinum, chromium, and other strategically critical elements. The remarkable continuity of the layering and individual horizons over hundreds of miles is impressive. Many staff members have visited specific areas and mines, but have not taken part in the disputes on the origins of this enormous magma body.[22]

Kimberlites

The varieties of volatile-rich (dominantly CO_2) rocks called kimberlite[23] have generated considerable interest because of their nodules and content of high-pressure minerals such as diamond, pyrope, and jadeite. The nodules have provided critical evidence for the depth of origin of the explosive diatremes. The temperature and pressure were estimated from the diopside solvus[24] and the Al_2O_3 content of enstatite[25] (see also Chapter 18). The unique composition of its minerals led to other experimental studies in addition to the phase behavior of the pyroxenes: micas, olivine, garnets, ilmenite, sulfides, perovskite and zircon. Because large volumes of rocks are processed in mining, sufficient zircons were recovered for age dating by G. L. Davis[26] (see Chapter 17 on geochronology).

The field investigations stemmed from determination of the *P–T* diagram for pyrope.[27] F. R. Boyd collected specimens at the Williamson Mine in Tanzania, Orapa in Botswana, the Lesotho pipes, the Premier mine and the dumps at Kimberley, South Africa, in 1973. Subsequently he traveled to Angola, East Griqualand, the Karroo, and Namibia. In preparation for the Second Kimberlite Conference, Boyd collected in Colorado and the Four Corners area. Four trips to Siberia included mines near the Arctic Circle and prospects further north. A week was spent at the Majhgawan mine in northern India, the kimberlite pipes of southeastern and western Australia, and eventually the Canary Islands. The chemical analysis of the minerals from these collections has resulted in major new ideas on the origin of kimberlites and the Earth's mantle. For example, the *P–T* data on the nodules exhibited an inflection in the geotherm (Figure 10.2) that Boyd and Nixon related to bodies of crystal–mush magmas interspersed in sheared lherzolites in the low-velocity zone.[28] The occurrence or absence of diamonds could be directly correlated with the depths of origin of the nodules. The kimberlite magma appears to have been derived from a garnet lherzolite source that was metasomatized by LREE-enriched CO_2 and H_2O vapors. From the thermobarometry of the inclusions in diamonds within the Kaapvaal craton in southern Africa, Boyd, J. J. Gurney, and S. H. Richardson concluded in 1985 that the diamonds were xenocrysts and had crystallized in a root extending to depths of 150–200 km.[29] The ambient temperatures at those depths

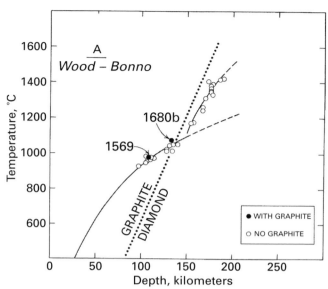

Figure 10.2 Estimates of the pressure (depth) and temperature of origin of nodules from the diamondiferous kimberlites in Lesotho and the Monastery Mine, South Africa, compared with the occurrence of graphite, using the diopside solvus and Al_2O_3 content of the enstatite. Data are compared with the shield geotherm of Clark and Ringwood. From F. R. Boyd, "A pyroxene geotherm," *Geochim. Cosmochim. Acta* 37 (1973), 2533–2546; S. P. Clark, Jr. and A. E. Ringwood, "Density distribution and constitution of the mantle," *Rev. Geophys.* 2 (1964), 35–88; Boyd and P. H. Nixon, "Origins of the ultramafic nodules from some kimberlites of Northern Lesotho and the Monastery Mine, South Africa," *Phys. Chem. Earth* 9 (1975), 431–454, at p. 434, figure 3a (reprinted with permission of Elsevier).

3000 million years ago were 900–1200 °C in accord with present-day heat flow.

Ocean-bottom sediments

Some 72 percent of the Earth is covered by ocean and the sediments on the bottom constitute the historical record of the ocean. It was important, therefore, to obtain an undisturbed vertical section. In the 1930s apparatus was designed by C. S. Piggot to obtain vertical cores up to 10 feet in length at depths of 2800 fathoms (16,800 feet).[30] Interest was motivated by the fact that the sediments in the ocean had a far greater concentration of radium than any other class of rocks on land.[31] The device for coring ocean bottom sediments consisted of a gun-fired sample tube and hoisting gear built at the Geophysical Laboratory. Some compaction of the core took place, but little distortion was seen.

The wealth of detailed information yielded by the cores included dating of the various horizons by noting the nonequilibrium between the radioactive elements. The method provides a means of measuring time over the past 300,000 years.[32] With knowledge of the dating, the rate of deposition of the sediments could then be estimated. The DTM used the cores to study the changes in orientation of the Earth's magnetic field. C. S. Piggot attributed the radium concentration to the increase in oxidation of ocean water with depth: the highest concentrations were found in red clays.[33] In general, the radium content was highest remote from land and in the greater depths. Cores were examined from the North Atlantic (see Figure 12.3), Cayman Trough, Gulf of Mexico, Grand Banks of Newfoundland, and the continental shelf off Ireland. Retrieval of the cores was a direct result of the high cooperation of the crews of the ships engaged.

Metamorphic facies

In preparing a list of potential geophysical investigations the original Advisory Committee suggested a dozen major topics relating to metamorphic rocks.[34] They were particularly concerned not only with the effects of differential stress and flowage, but also with solution and recrystallization "in imitation and elucidation of the natural alteration of minerals."[35] The initial investigations were based on the substantial field studies of internationally recognized metamorphic geologists. In 1913 J. Johnston was joined by Paul Niggli, a visiting investigator from Switzerland, to prepare a paper on "The general principles underlying metamorphic processes."[36] They enumerated the effective factors – temperature, uniform pressure, stress (nonuniform pressure), gross composition of the system – and called attention to the influence of the rate of reaction. Although A. F. Buddington only spent fifteen months at the Laboratory, he spent the two field seasons mapping in the Adirondacks. Eventually Buddington put in over twenty-two field seasons there.[37] A research associate, Pentti Eskola from Finland, described the contact phenomena between gneiss and limestone in western Massachusetts in 1922.[38] He distinguished the wollastonite, diopside, tremolite, and quartz limestones and attributed the sequence to successively lower temperatures. The sequence was expanded by N. L. Bowen in 1940 for siliceous dolomites to larnite, merwinite, spurrite, akermanite, monticellite, wollastonite, periclase, diopside, forsterite, and tremolite.[39] Although his forecast was most appropriate and he "hoped the equilibrium curves of the different silicates with carbonates may soon be determined experimentally," it would be almost forty years before fulfillment. The next major field study was by visiting investigator T. F. W. Barth from Norway who dealt with the petrology and metamorphism of the Paleozoic rocks of Dutchess County, New York. This classic study, published in 1936, demonstrated that argillaceous sediments recrystallized sequentially into slate, phyllite, schist, and gneiss during a period of orogenesis.[40] Because the sediments were subjected to prolonged heating in liquids of magmatic or palingenic origin, he proposed to use the term "migmatic" to distinguish them from rocks without anatectic or metasomatic alteration.

In order to test his experimental results on the role of water in metamorphism and potential reconfiguration of the facies assemblages, H. S. Yoder visited many of the classic metamorphic areas, usually under the guidance of the original mappers.[41] Professor C. E. Tilley of Cambridge showed him the details of the progressive metamorphism of Barrow's zones, and the definition of "isograd" was debated.[42] The eclogites and their conversion to hornblende schists were studied at Glenelg, Scotland by A. R. Alderman in the 1930s.[43]

Professor T. C. Phemister pointed out the key outcrops of the metamorphism in the Stonehaven-Aberdeen region of Scotland where a series of impure feldspathic grits and sandstones interspersed with shales and calcareous beds resulted in high-grade schists and gneisses, and eventually developed into a migmatite complex in its northern part.[44] The rocks of the Stavanger region of Norway, on which Goldschmidt based his classic 1921 account of injection metamorphism and the evolution of migmatites,[45] were examined. The metamorphic rocks of the Trondheim region of Norway were examined in the company of Professor Th. Vogt and those in the Oslo region with the help of Professor Barth. Through special circumstances the famous Orijarvi region of Finland was toured with both Professor Pentti Eskola and Professor H. Tuominen.[46] The vigorous debates on the alleged isochemical nature of the facies classification as espoused by Eskola and the important role of water[47] as well as structural controls[48] were most illuminating. A review of the debate was eventually given by Barth in 1962 in which he accepted the relationship between the chemical activity of H_2O and the facies, and emphasized that there is a regular relationship by virtue of the interdependence of the variable parameters in the geological process.[49] After an examination of the isograd problems in metamorphosed iron-rich sediments, Yoder[50] toured the iron formation in Biwabik, Michigan, under the guidance of Dr. H. L. James.[51] The purpose was to test the concept of appearance or, preferably, the disappearance of a mineral as an isograd marker, as emphasized by Bowen in 1940.[52] It is evident that any experimental result should be tested in the field with full access to the array of natural circumstances.

The regionally metamorphosed rocks of central-western New Hampshire provided recumbent folds and gneissic domes.[53] A detailed study of the composition of the coexisting minerals there was made by D. Rumble[54] that led to the removal of some uncertainty in the petrogenetic grid based on Fe-Mg partitioning. He also focused on the Fe-Ti oxide minerals in an effort to determine the conditions of equilibrium or disequilibrium with respect to the volatile components.[55] That led to his considering the chemical potential gradients between sedimentary beds.[56] He concluded that neither H_2O nor

O_2 was perfectly mobile and that the circulation of the fluid was restricted sufficiently so that the local mineral assemblage controlled the properties of the fluid. After examining the oxygen isotope geochemistry, he believed that although isotopic equilibrium had been obtained, the mineral assemblages were not in equilibrium with the fluid.[57] In short, the rocks were not metamorphosed in the presence of a fluid of uniform chemical or isotopic composition. From the oxygen isotopes Rumble and T. C. Hoering found that permeability and intensity of devolatilization are correlated variables.[58]

In a continuing study of the role of fluids in regional metamorphism, Rumble found calc-silicate rocks equilibrated in their oxygen isotope composition with a pre-metamorphic quartz monzonite dike and other contiguous rocks via a pervasive fluid phase.[59] Both reaction progress and oxygen isotope data showed that fluid–rock ratios of 4:1 by volume were required to achieve the observed equilibration between different rock types. These same rocks contained a rare occurrence of Devonian brachiopods replaced by wollastonite.[60] A regional scale study of the central New Hampshire metamorphic terrane revealed a widespread metamorphic fluid flow system with pervasive hydrothermal graphite in microscopic veinlets as well as in outcrop-scale vein systems.[61] C. P. Chamberlain and Rumble showed that one of the graphite vein systems was located in a granulite facies metamorphic hot-spot and speculated that the thermal anomaly may have been affected by fluid flow.[62] With the discovery of record low $\delta^{18}O$ rocks from the coesite-eclogite facies of east-central China by Yui and co-workers[63] came the recognition that an entire geothermal system had been subducted *en masse* during continental collision of the North and South China tectonic plates.[64] Further study showed that the age of the geothermal system was Neoproterozoic and the observed negative $\delta^{18}O$ values were acquired from meteoric water reflecting a cold climate that may be related to a "Snowball Earth."[65]

Ecological provinces

The unique array of field studies of M. L. Fogel has covered a wide spectrum of problems that bear on the management of our resources in relation to predictions of climate change. With the use of advanced biochemical, isotope, and spectroscopic techniques, she investigated how biochemicals survive, transform into other molecules, and become part of the carbon reservoir. In 1981 and 1982 the biogeochemistry of the stable nitrogen isotopes in the hydrothermal areas of Yellowstone National Park were studied to discover the factors that govern the $\delta^{15}N$ of organic nitrogen in evolutionarily primitive organisms.[66] The marshes of the Delaware estuary were sampled both for bottom sediments and suspended particulate matter and the seasonal variability in stable carbon and nitrogen isotope ratios determined. It was found that biogeochemical processes influenced the isotopic distributions in the estuary to a greater extent than physical mixing.[67] Further studies in the estuary revealed that the cycling of NH_4^+ was mainly due to bacterial nitrification, assimilation by primary producers in the spring and microbial regeneration in the summer.[68] Samples were also collected from three anoxic marine basins: Black Sea; Saanich Inlet, British Columbia, Canada; and Framvaren Fjord, Norway. In each basin the $\delta^{15}N\text{-}NH_4^+$ was greatest near the O_2/H_2S interface.[69] G. E. Bebout and M. L. Fogel studied the metasedimentary rocks in the Catalina Schist, California, subduction-zone metamorphic complex for changes in nitrogen concentration.[70] They found a decrease in concentration and an increase in $\delta^{15}N$ with increasing metamorphic grade. The distillation–devolatilization process implicated in the study may govern the behavior of other trace elements partitioned into hydrous fluids during devolatilization. The possibility of using nitrogen as a tracer in large-scale volatile transport was recognized.

An unusual opportunity became available to M. L. Fogel and colleagues when Hurricane Gordon interrupted their sampling of a coastal transect off Beaufort, North Carolina in 1994.[71] Immediately after the storm's passage the transect was resampled and the significant changes in primary production in the continental shelf zone recorded. Significant increases in chlorophyll *a* CO_2 fixation, and bacterial production were found over relatively large areas. Stable isotope compositions of suspended particulate material shifted quickly and recorded the biological perturbations to the water column. They concluded that the primary production associated with major storms may influence the sedimentary organic carbon in coastal areas.

Because of the role of nitrogen in limiting the primary production in coastal waters, the input of nitrogen via rain events was evaluated in generating phytoplankton.[72] These events led to nuisance blooms of microalgae, which are often inedible and give rise to hypoxia and anoxia. By using field-based, high-pressure liquid chromatography, it was possible to characterize the phytoplankton production and the species composition responses to the atmosphere contribution of nitrogen.[73] The coastal zone has only recently been recognized for eutrophication found in nutrient-impacted freshwater ecosystems. An entirely different ecosystem was studied by Fogel and G. Miller who took advantage of the extreme conditions in the Australian Outback.[74] They were particularly concerned about how ancient humans had altered the climate by large-scale brush burning. The isotopic signatures in fossils, e.g. emu eggshells, were used to reconstruct the ancient environments.

P. E. Hare examined the early sites of human habitation in Murray Springs (Arizona), Sunnyvale (California), and others for the purpose of determining the age of the bones with the racemization method (see Chapter 11).[75] The results were reasonable from a geological viewpoint, but were not acceptable in the opinion of Hare's sponsors, the Seventh-day Adventist Church.[76]

In addition to the above described field studies, practically all staff members visited specific localities to obtain rocks, minerals, or fossils to be used in specific experiments. The staff of the US Geological Survey and members of the nearby university staff were always helpful in locating the most useful outcrops.

For many years, the younger staff members and fellows of the Geophysical Laboratory accompanied Dr. J. Frank Schairer in the summer to problem areas being actively mapped by others. The physical hardships of climbing, back-packing and camping generated an *esprit de corps* of cooperation and companionship that contributed immensely to the friendly and helpful spirit of the Laboratory. The campfire discussions brought a broad measure of understanding of the field problems and often gave rise to the outline of critical experiments to test the ideas presented by the field geologists. These experiences provided a focus for the Laboratory on the important and critical problems of geology.

BIOGEOCHEMISTRY

The application of organic chemistry to geological problems[1] at the Geophysical Laboratory arose out of the research interests of the newly arrived director, Philip H. Abelson, in 1953.[2] He said he was stimulated by a conference on "Comparative Biochemistry, Paleoecology, and Evolution" organized by W. P. Woodring.[3] The plethora of opportunities for research have led to investigations of the organic constituents of fossils, geochronology and geothermometry of fossils, organic mineral deposits, biomineralization, hydrologic cycles, ecology, paleoclimatology, primitive atmospheres, diagenesis, physiology of fossils and humans, archeology, isotopic fractionation in organic compounds, the survival limits of thermophylic and barophylic organisms, and the many aspects of astrobiology. No doubt the Geophysical Laboratory will play a pioneering role in these fast growing areas of science, evolving into new areas as other organizations fulfill maturing directions.

Organic constituents of fossils

Organic material equivalent in quantity to the weight of the Earth has been created by living creatures since life originated on this planet. Almost all of this has been metabolized by other creatures, with a few notable exceptions. Under special conditions of deposition and burial, the organic matter represented by coal, oil, and oil shales has been preserved. Some of the chemicals found in these fuels are only moderately altered products of compounds originally part of living organisms. Examinations of a variety of fossil shells and bones have now revealed that small but important quantities of organic materials are often preserved within these fossils.[4]

In the mid-1950s Abelson used paper chromatography to demonstrate that fossils as old as 360 million years retained amino acids from some of their original proteins.[5] Abelson also determined that the breakdown of amino acids in fossils could be simulated in the laboratory by substituting elevated temperatures for geological time, thereby demonstrating their potential as stratigraphic markers and geochronometric tools. These "chemical fossils" complemented the classical methods of paleontology. The book *Biochemistry of Amino Acids*, edited by P. E. Hare, T. C. Hoering, and K. King, Jr. (1980), has been the definitive work on the subject.[6]

Geochronology

Fossils are usually arranged chronologically by the sequence of layers in which they are found, by the evolutionary changes in their morphology, and by the dates assigned to strata obtained by other methods, e.g. isotopic decay schemes. Hare discovered a new method wherein the racemization of amino acids was a function of time and could be used for dating.[7] All the amino acids in proteins exist in two configurations that are mirror images, or optical isomers, designated D and L. Hare and Hoering were successful in separating these isomers with gas chromatography and high-pressure liquid chromatography.[8] They learned that biologically produced amino acids that are dominantly L, transformed spontaneously to D abiologically as a function of time (i.e. racemized). Hare developed these observations into a method for dating fossils as old as 20 million years.[9]

From the early paper chromatography, the techniques have evolved to very-high-resolution capillary gas chromatography, high-pressure liquid chromatography and eventually to a combination of gas-chromatographic and mass spectrometric methods. A field-portable, liquid chromatograph was made by Hare and G. H. Miller to measure amino stratigraphic sections on site.[10]

Geothermometry

After demonstrating that the amino acids extracted from fossils were original material and not modern contamination, the question of their stability through temperature changes was raised. By heating the amino acid alanine, it was shown by Abelson in 1956 that it might persist for millions of years if the temperature remained below 60 °C (Figure 11.1).[11] Conversely, if the age were known, then racemization of the amino acids could be used for determining paleotemperatures. For example, Engel and colleagues found the average temperature experienced by a Sequoia tree over the past 2185 years.[12] The difference in D/L of Pleistocene fossil molluscs from California, Washington, and Florida were attributed to temperature differences by J. F. Wehmiller and D. F. Belknap in 1978.[13] Another team of researchers noticed that the amino acids in bristlecone pine of the same age, collected at different elevations exhibited a difference of 6 °C mean annual temperature because of the difference of 650 m.[14] If amino acids are found in a fossil, an upper limit can be set on the temperature to which the fossil has been exposed. This in turn can be related by means of geothermal gradients to an estimate of burial history. It appears that accurate and specific paleotemperatures can be ascertained if the preserving environments are taken into account.

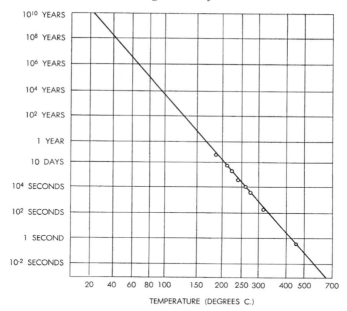

Figure 11.1 Persistance of alanine at various temperatures. The vertical component indicates the length of time it took for 63 percent of the alanine sample to break down. *Source*: From P. H. Abelson, "Paleobiochemistry," *Scientific American* 195 (1956), 92.

Stable isotopes

A powerful technique for illustrating biological change in geological materials was acquired when Thomas C. Hoering joined the staff in 1959. As a result of his skills in mass spectrometry, the Geophysical Laboratory developed dedicated facilities for measuring the stable isotopes of C, H, O, N, and S.

When inorganic carbon is converted into living matter the light isotope of carbon tends to be preferentially fixed. Organic carbon thus has a lower C^{13}/C^{12} ratio than the carbonate or carbon dioxide of the inorganic environment.[15] P. H. Abelson and Hoering demonstrated this effect experimentally in 1960 by culturing algae, hydrolyzing the protein found, separating pure amino acids by ion-exchange chromotography, combusting a portion of the amino acids to carbon dioxide, decarboxylating a portion of the amino acids with ninhydrin and purifying the liberated carbon dioxide, and finally performing isotopic analysis on the carbon dioxide with mass spectrometry.[16] The process was monitored in some algae samples by following ^{14}C radioactivity to be sure complete recovery of the fractions of amino acids was obtained. M. Estep (Fogel) found that present-day blue-green algae and bacteria growing in CO_2-rich hot springs exhibited the same depletion in ^{13}C as in Precambrian stromatolites formed by the same types of organisms.[17] She concluded that the atmosphere in Precambrian times was, therefore, probably enriched in CO_2 by several percent relative to the present-day atmosphere.

In an unprecedented collaboration between the Geophysical Laboratory and the Carnegie Institution of Washington's Department of Plant Biology at Stanford, M. L. Fogel and Joseph Berry solved one of the stable-isotope geochemistry's oldest problems, the "Dole Effect,"[18] that was identified in 1936. Atmospheric oxygen is anomalously enriched in the heavy isotope, ^{18}O, and previous studies could not account for this effect. Fogel and Berry discovered that a large isotope fractionation occurred during the uptake of O_2 in photorespiration, a process that accompanies the photosynthesis reaction, and thus they could account for the isotope enrichment.[19] It appears that photosynthetic oxygen originates from water, not from CO_2.

Many studies focused on the fractionation of stable nitrogen isotopes in organic matter. "Cellular nitrogen metabolism is complex, and nitrogen isotope fractionations occur during assimilation, synthesis, transfer, and excretion of nitrogenous compounds."[20] Bacteria were grown on a single amino acid as a source of nitrogen. The cells grown on glutamic acid were enriched in ^{15}N relative to substrate, whereas those grown on alanine were depleted in ^{15}N relative to the source. They concluded that the isotopic compositions of micro-organisms are associated with the metabolic pathways in their synthesis. Whereas fossil food webs may be established, care must be taken to resolve post-depositional amino acids from indigenous ones. To evaluate those effects, Hare and colleagues examined the bones of pigs raised domestically on diets with known compositions as well as several herbivores and carnivores from modern and ancient environments.[21] They found that many animals have similar patterns in composite isotopic labeling even though they varied in feeding habits, but exceptions were observed where the essential amino acids were controlled by an individual animal's metabolism. In another study, by G. E. Bebout and Fogel in 1992, the nitrogen isotope compositions of metasedimentary rocks in the Catalina Schist of California indicated that N could be used as a tracer of large-scale volatile transport.[22]

A method for determining the isotopic composition of elemental sulfur became a possibility with the development of an extraction scheme for elemental sulfur and a bromine pentafluoride (BrF_5) technique. This made it possible to study the sulfur isotopes in a sediment core taken from a coastal salt marsh.[23] It was found that the ^{34}S content generally decreased with depth and was on average greater than that in the coexisting pyrite phase. Apparently isotopically lighter ^{32}S sulfide could be selectively incorporated into pyrite, whereas isotopically enriched ^{34}S sulfide was converted into elemental sulfur. The new procedure along with the analysis of the other sulfur phases (e.g. pyrite, iron monosulfides, organic sulfur, sulfite, and sulfide) will no doubt be helpful in elucidating the sulfur cycle in marine sediments being actively studied.

Marine environments

As a result of purported "epidemics" of phytoplankton blooms, attention has been directed toward bioassays of the production of organic materials in coastal waters. Although only 10 percent of the surface area of the world's oceans account for 40 percent of the oceanic production, it is suspected that this is due to anthropogenic nutrient loading. The seasonal variability of suspended particulate matter was studied by L. A. Cifuentes and colleagues in the 1980s in the Delaware estuary using the isotopes of carbon and nitrogen.[24] They found that biochemical processes influenced isotopic distributions in the estuary to a greater extent than physical mixing. There were clear differences in suspended particulate matter and bottom sediments, which could be remineralized in the late summer and fall to support additional primary production. Their results were used to identify this algal production from the alleged high proportion of terrestrial or sewage component.[25] In 1994 H. W. Paerl and Fogel attributed a significant fraction of the external nitrogen loading to atmospheric deposition by measuring the coastal rainfall events over a two-year period.[26] Earlier they had estimated that >50 percent of the "new" nitrogen input to coastal waters was from atmospheric deposition.[27] They noted that the rising frequencies and magnitudes of coastal algal blooms may be attributed to this important source.[28] Predictable changes were observed in the isotope ratios of the different pools of carbon and nitrogen in the anoxic basin of the Framvaren Fjord, Norway.[29] The isotopic distribution recorded the anoxic conditions with some variations attributed to the inputs to the sediments relative to the in situ production. The investigators noted it was difficult to separate out these sources without compound specific analysis. After additional studies in two other anoxic basins (the Black Sea and Saanich Inlet in British Columbia, Canada), D. J. Velinsky and co-workers suggested that NH_4^+ utilization is the dominant process controlling the nitrogen signature in the water column.[30]

Organic sediments

About 95 percent of the world's organic matter is in the form of a complex, insoluble material called kerogen.[31] It is believed to be the probable source of petroleum. In the 1960s Abelson set out to find out how kerogen is formed and determine the processes involved in the origin of petroleum.[32] In addition, the data would no doubt contribute to an understanding of the evolution of life. The first problem that had to be overcome was isolating the kerogen from its associated minerals (e.g. pyrite). The extraction techniques and subsequent isolation of its various constituents (e.g. fatty acids, amino acids, carbohydrates, and porphyrins) no doubt all affected their composition. According to Abelson, "The biosynthetic activities of living matter are the ultimate source of kerogen."[33] The familiar lipids, carbohydrates, and proteins of micro-organisms in anaerobic sediments are not similar to kerogen chemically and therefore behave differently. The focus at the Geophysical Laboratory was (1) to define the structures in kerogen, and (2) determine the mechanics of its formation.

In following the fate of nitrogen, Abelson found that most of it became insoluble, not by biological attack, but by chemical combination. With hydrogenation at elevated temperatures (~380 °C) some pyrolysis takes place, so Abelson used anhydrous hydrogen iodide to break up the linkages at a lower temperature (~250 °C). In the Green River shale, he was able to split C–C bonds and produce an array of by-products identifiable by gas-liquid chromatography. After an array of experiments on different shales at a range of temperatures, Hoering and Abelson concluded that "mild thermal degradation of kerogen is the principal mechanism by which hydrocarbons in natural gas and petroleum are produced."[34] A year later, in 1964, they reported a significant difference in the aromatic or hydroaromatic structures in coal compared to the aliphatic carbon residues in young kerogen.[35] Even billion-year old organic matter from the Nonesuch Shale of Michigan provided evidence that the biochemical processes were typical of sediments of all ages. One of Hoering's concerns was that the older sedimentary rocks may have been contaminated by the migration of petroleum fluids.[36] By measuring the carbon isotopes, he found no differences in the extractable and insoluble organic matter in the rocks, but did caution that other criteria will need to be employed.

Because saturated fatty acids are prominent constituents of living organisms and have intrinsic chemical stability, Hoering and Abelson improved a procedure for isolating long-chained fatty acids from a mixture of organic acids.[37] Preliminary results on rocks of Paleozoic age or younger indicate that the amounts of fatty acids decrease with age and metamorphism, with the relative amounts of odd-numbered fatty acids in carbon numbers increasing until there is a relatively smooth distribution with little predominance of even-numbered acids that predominate in the extractable acids. Five years later, in 1970, Abelson and Hare made the chance observation that amino acids were not recovered in tracer experiments with the Green River shale.[38] In addition, amino acid tracers disappeared and there appeared to be irreversible reactions of kerogen and humic acid. Their experiments showed that kerogen is an effective scavenger for amino acids and might be expected to reduce the level of many free amino acids in the natural environment within a relatively short period of time.

A search for molecular fossils in the kerogen of eighteen Precambrian sedimentary rocks was made by Hoering and V. Navale in the 1980s.[39] Samples older than 1.6 billion years yielded no detectable hydrocarbons. It was hoped that better preserved samples that did not suffer higher surface temperatures may increase the probability of learning about the evolution of Precambrian life.

One of the types of kerogen related to vascular plants, the vitrinites, was studied by G. D. Cody and colleagues in the late 1990s because of its potential for recording the thermal history of associated rock with diagenesis.[40] The thermal history is crucial in oil exploration as well as in providing constraints on paleoheat flow and geodynamic burial and uplift histories. They found a correlation of luminescence alteration with thermal maturity that depended on kinetics and the chemistry of photo-oxidation. But to understand thoroughly the reactions they recommended a systematic study of simple compounds that were specific to different vitrinites.

In 1999, high-level quantum mechanical calculations were brought to bear on the problem of kerogen evolution.[41] They were able to prove that the macromolecular structure of type III kerogens must have undergone complete molecular rearrangement through chain bond interchange even at relatively low thermal maturity. Thus, while the majority of the macromolecule remains intact during diagenesis, the initial biomacromolecular structure is lost.

In 2002 a team of investigators demonstrated that the chemical differentiation related to the nanostructure of vascular plant cell walls is preserved in ~400 Ma cherts from the Rhynie locality in Aberdeenshire, Scotland.[42] To do this they used a unique scanning transmission (soft) X-ray microscope (STXM) located at the National Synchrotron Light Source, Brookhaven National Laboratory. Their results were significant in that such chemical differentiation could only be explained by the ancient plant fossil (Asteroxylon) having, while alive, the capacity to synthesize the biopolymer lignin. As such, this constitutes one of the first direct assessments of biopolymeric assemblages in long-extinct fossilized life. Subsequent work with plant fossils lower in the phylogeny suggests that Asteroxylon may have been the first plant to synthesize lignin.[43]

Biomineralization

The relative scarcity of Precambrian fossils has been attributed to their lack of preservable hard parts. On the other hand, Cambrian rocks contain a variety of organisms that possessed hard parts of chitin, calcium carbonate, calcium phosphate, and silica. The $CaCO_3$-secreting organisms in particular have increased in number and variety since early Cambrian time.[44] Calcium carbonate as calcite, aragonite, or a mixture of the two is intimately associated with the organic matrix, which is no doubt involved in the shell-forming process. The structures are classified into a few basic types:[45] brick-wall pattern, prismatic, crossed lamellar, foliated, and irregular. The organic content of each structure varies and may range from 0.01 to 5 percent, and the amino acid contents are similar whenever the morphology and shell structure are closely related. Hare and Abelson found that chitin may have preceded minerals as the principal hard part.[46] Arthropod evolution, in contrast to calcified organisms of chitin, appears to have remained dominant and calcification was subordinate.

One of the few naturally occurring crystalline organic substances[47] found in sedimentary formations is fichtelite, with three major structural components. After detailed study of material from the type locality, Hoering was able to describe the structures of the components along with retene, iosine, and abietic acid.[48] Elemental sulfur converts fichtelite and abietic acid to retene. He searched other localities but had negative results.

The mobilization and deposition of carbon as graphite in veins intrigued D. Rumble and Hoering.[49] The rare graphite veins in New Hampshire that occur in sillimanite-grade metasedimentary rocks have been exploited commercially. It was first proposed that the carbon is mobilized by metamorphism of common sedimentary rocks. Metamorphism of shales containing reduced organic matter produces aqueous fluids with $CH_4 > CO_2$ at low fugacity of O_2. On the other hand, metamorphism of argillaceious limestones give rise to fluids with $CO_2 > CH_4$. Graphite precipitates when the two aqueous fluids of different CO_2/CH_4 ratios are mixed.[50] For example, if fluids A and B in Figure 11.2 are mixed, carbon as graphite is precipitated and could be propagated through rocks by the mechanism of hydraulic fracture. This new mixing hypothesis is supported by data on the isotopes of C.[51]

Paleodiets

The food webs in ecosystems where a number of different organisms are interacting have been determined with isotopic analyses. In 1980 Estep (Fogel) and H. Dabrowski reported how they were able to trace the food chains of a snail that had been feeding on a specific alga with the stable isotopes of hydrogen.[52] Whereas animals feeding on a particular diet fractionate the hydrogen isotopes, the fractionation is small and the animal tissues closely resemble those of their food.[53] In short, they were able to confirm the old adage, "You are what you eat!" They noted that the isotopic composition of the water ingested had no significant effect. Although some differences were

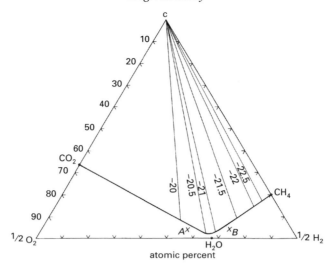

Figure 11.2 Phase diagram for C-O-H at 600 °C and 4 kbar. The tie lines from C (graphite) to the heavy curve marking coexisting fluid compositions are labeled with the calculated fugacity of oxygen. Mixing of fluids A and B, for example, results in the formation of graphite. From D. Rumble and T. C. Hoering, "Carbon isotope geochemistry of graphite vein deposits from New Hampshire, USA," *Geochim. Cosmochim. Acta* 50 (1986), 1246, figure 6. Reprinted with permission of Elsevier.

obtained between modern and fossil whale bones, N. Tuross and colleagues found that the food sources could be identified.[54] In general, no evidence was found for the alteration of the carbon and nitrogen isotope composition of weathered bone collagen.[55] Even the eggshells of the African ostrich, ubiquitous in archaeological sites, retain their organic matrix sufficiently well to estimate their age by protein hydrolysis, amino acid epimerization and racemization, and amino acid decomposition reactions.[56] Their techniques provided accurate ages to within 10–15 percent for strata deposited within the last 200,000 years in the tropics and the last million years in colder regions such as China. In experiments on laboratory-reared pigs by Hare and colleagues,[57] it was demonstrated that in controlled diets the composition of C_3 or C_4 plants could easily be distinguished. The most interesting aspect was that the isotopic compositions could be traced from muscle tissue, to collagen, to bone. In a study of African elephants, it was shown that with time there had been a shift in diet from trees and shrubs to grass. Changes in climate, fire frequency, and expanded foraging were suggested as possible causes for this change in diet.[58] The expansion of C_4 grasses in Central Australia during the period 45,000–65,000 years ago was deduced from the carbon isotope in the fossil emu eggshells.[59] From those data it was inferred that the relative effectiveness of the Australian summer monsoon was greatest at that time, least effective during the last glacial maximum, and moderately effective during the Holocene. The Pleistocene cave fauna of southern England revealed the dietary discrimination of herbivores and carnivores.[60] Distinct differences were also recorded in the isotopic composition of the tooth enamel of the herbivores and carnivores, which were similar to those found in modern specimens from similar climatic environments.

The reconstruction of paleodiets for humans was the goal of Fogel and her team in the 1990s.[61] The establishment of agriculture in North America was determined by comparing the hunter-gatherers with the plains agriculturalists who cultivated corn and augmented their diets with buffalo meat. The analysis of the diets of the Easter Islanders revealed a fascinating history as they changed from fish to chickens and rats. The investigators predicted major strides in studying human evolution via biogeochemical tracers back to the time when anatomically modern humans evolved. The wooden artifacts buried at archaeological sites are usually degraded, but Filley and colleagues were able to identify the coffin in the King Midas tomb at Gordion, Turkey, from other wood materials by examining the cycling of the nitrogen isotopes.[62]

Special note must be made of the unusual opportunity presented by the botanical collections made on the Lewis and Clark expedition between 1804 and 1806. The carbon isotopes were measured on eleven specimens representing the earliest plant collection predating the Industrial Revolution. An investigation in 2002 showed that the herbarium specimens could provide data for reconstruction of the temperature, climatological conditions, and atmospheric chemistry for a time before records were kept.[63] This unique untapped reservoir of information provides a new benefit from museum specimens as yet unrealized.

Barophyllic and thermophyllic organisms

The discovery of abundant life at hydrothermal vents on the sea floor[64] gave rise to a new view on the limits of survival of organisms and expanded the range of conditions under which life may have originated. Estep (Fogel) studied the algae in the hot springs of Yellowstone National Park and found that the hydrogen isotopic composition could be related to temperatures up to 85 °C.[65] She observed that the similarities of modern hot spring algae and Precambrian stromatolites were numerous. According to J. A. Baross, life may be sustainable to at least 115 °C.[66]

Although the pressure at the greatest ocean depth is about 1 kbar, microbial activity has been demonstrated at pressures up to 16 kbar.[67] Pressure expands the thermal stability of aqueous phases and enhances the variety of organic reactions. In the case of pyruvic acid, increased yields of cyclic aromatic compounds were obtained at 2 and 5 kbar.[68] In addition, studies of the stability and reaction pathways of citric acid were extended to 300 °C under the same range of pressures.

It appears that the surprising high-pressure and high-temperature stability of some microbes provides for a wider range of plausible environments for the origin of life (see Chapter 14 on astrobiology).

12

GEOPHYSICS

Although the name of the Geophysical Laboratory implies that a large component of the work would involve geophysics, the classical fields of endeavor now included under the term have played a small role over the years. For brief periods, however, gravity, heat flow, electrical conductivity, thermal conductivity, density, magnetism, tectonophysics, oceanography, and seismology, have all been investigated. Some of these studies were undertaken in cooperation with the Department of Terrestrial Magnetism, which was established in 1903. It initially dominated the field of magnetism, complementary to the national bureaus, and was a pioneer in explosion seismology after World War II.

Gravity

One of the many interests of F. E. Wright was the difference in gravity between the Earth and Moon and the resulting differences in geomorphology and isostatic compensation.[1] Pursuing this interest, he persuaded F. A. Vening Meinesz in 1928 to install his pendulum for making gravity determinations at sea on US submarine S-21.[2] From subsequent measurements, they concluded that some oceanic deeps were uncompensated whereas the Mississippi delta was practically compensated in spite of the enormous load of sediment laid down each year. Inspired by the facility of occupying a large number of stations at sea, Wright and J. L. England developed an improved torsion gravity meter mounted on a truck so that twenty or more stations could be occupied in a day.[3] Although most of the stations occupied were in the eastern USA, the apparatus was set up in 1940 on the active volcanoes of Santa Maria and Santiaguito in Guatemala to assess the changes in the magma chamber in conjunction with other geophysical measurements.[4]

A gravity program with closely spaced stations was initiated along the eastern part of the Allegheny Mountains to the coast. Unfortunately, the program was held in abeyance during World War II and not continued.[5] In Ajo, Arizona, P. M. Bell and R. F. Roy undertook gravity measurements in the 1960s to account for the abnormally high heat flow.[6] They found the heat flow to be about 10 percent higher than the regional mean and presumed it resulted from a thick layer of low-conductivity alluvium.

Heat flow

An early attempt (1912) to measure the thermal gradient in the crust was made by J. Johnston and L. H. Adams.[7] They used both mercury thermometers and an electrical resistance thermometer in wells as deep as 5230 feet near Charleston, West Virginia. It was believed that such measurements might also have economic importance in identifying layers rich in coal or oil, indicated by a higher temperature gradient. They were very adamant that the resistance thermometer was "ten times as good" as the mercury thermometer. This conclusion no doubt stemmed from Johnston's experience in Ohio where three mercury maximum-reading thermometers were lowered in a copper tube down to 3000 feet.[8] A discontinuity in the temperature–depth curve was noted at about 1000 feet, where gas appears in the well. The lower temperatures above 1000 feet were apparently the result of circulating air, and not representative of the temperature of the sedimentary rocks.

An opportunity for measuring heat flow in long tunnels occurred in the construction of the Arlberg and Taverin tunnels in Austria. With the underground temperature observations and the laboratory measurement of thermal conductivity of the various rocks, S. P. Clark, Jr. found that relatively high geothermal fluxes extend into the eastern Alps.[9] Subsequently, the relatively high heat flow (2.2 mcal/cm^2 s) in Arizona was investigated by Bell and Roy, who related the results to the gravity and seismology data of the region (see above).[10]

Geotherm

The problem of the cooling of a primitive Earth was investigated by L. H. Adams[11] after the revealing calculations of A. Holmes[12] in 1916 and of H. Jeffreys in 1924.[13] Adams generated a "most probable" geotherm down to 300 km, assuming the age of the Earth was 1.6 billion years, the generation of radioactive heat was constant, and the Earth was covered with a substantial molten layer. In the late 1940s, W. D. Urry also calculated the geotherms for the Earth at various times taking into account the exponential decay of radioactive elements and the variation of surface heat flow with time.[14] Later, in 1961, Clark derived geotherms as a function of time for various models calculated with the aid of a digital computer.[15] He concluded that the distribution of radioactivity could not be inferred from the near surface heat flow. Furthermore, the variability of heat flow cannot be attributed to different degrees of concentration of radioactivity in the outer few hundred kilometers of an initially homogeneous Earth. An innovative approach to the geotherms was found by F. R. Boyd, Jr. in 1973, who applied the pyroxene geobarometer and geothermometer.[16] On the basis of the composition of coexisting pyroxenes in nodules from kimberlites, he obtained a quantitative measure of the geotherm. In general, the geotherm derived from the nodules substantiates the geophysical estimates based on surface heat flow. His results also showed that cratons have been cool relative to oceanic plates for at least 3 billion years. One surprising result was an inflection in the geotherm under northern Lesotho that may be attributed to a region of partial melting.

Tectonophysics

Prior to the opening of the Geophysical Laboratory and during its formative period, grants were made to F. D. Adams in 1903–1905 by CIW to study the flow of rocks. A 120-ton press was set up at McGill University to investigate the deformation of marble, granite, diabase, and other rock types at temperatures up to 1000 °C. One of the goals was to understand the origins of crystalline schists, a subject of special concern to Van Hise who served on the CIW Advisory Committee in Geophysics. Attempts were made by E. D. Williamson in 1917 to determine the effects of strain on heterogeneous equilibrium; that is, on a mineral assemblage and a liquid.[17] He agreed with F. E. Wright and J. C. Hostetter[18] in regard to a free surface, but was doubtful that equilibrium could be obtained on a thrust surface. A theoretical interpretation of the flow in stressed solids by R. W. Goranson in 1940,[19] using thermodynamic potential relations for different physical conditions, was corroborated by experimental studies. Those experiments were not published, presumably because of Goranson's assignments during World War II, but support for his theory was provided by the more detailed laboratory studies of D. Griggs at Harvard.[20] For a five-year period, two postdoctoral fellows from Yale undertook a field and experimental program relating the conditions of flow to the resultant plastic strain. Displacement of slip folds, flexures in lavas, detachment structures in sediments, and theories for drag folds were focused on by E. Hansen and W. H. Scott. On the basis of these and other studies, Hansen developed the concept of "strain facies" and recorded his ideas in a book published in 1968.[21] Goranson was dubious, however, and questioned direct comparison of rock flow in nature with those obtainable in the laboratory.

The hypothesis of melting at stress dislocations in the Earth was examined by P. M. Bell and H-k. Mao in the early 1970s.[22] The theory they proposed included mechanisms for generating two or more igneous liquids that are not necessarily related to the parent rock by fractionation trends. Diffusion trends in stress gradients may predominate. The process could result in the formation of incompatible multiple lavas, which are commonly observed in single flows.[23]

Figure 12.1 Apparatus for deep-ocean bottom sampling with canister for explosive charge to drive core casing into sediment. From C. S. Piggott, "Core samples of the ocean bottom and their significance," *Sci. Monthly* 47 (1938), p. 212, figure 4. By permission of the American Association for the Advancement of Science.

Oceanography

The discovery of the high radium content of ocean bottom samples collected by the auxiliary brigantine *Carnegie* led C. S. Piggot to develop a device for coring ocean bottom sediments in 1935 (Figure 12.1).[24] Using a gun-fired sample tube and hoisting gear built at the Geophysical Laboratory, he obtained cores with the cooperation of the crews on the ship *Atlantis* (Woods Hole Oceanographic Institution). Cores up to ten feet in length were recovered from depths as great as 2700 fathoms (Figure 12.2). Piggot's goal was to obtain the historical record of past life at the ocean bottom as opposed to "grab samples" taken from the very surface of the bottom. The fourteen cores recovered were obtained during May 1936 between the Grand Banks of Newfoundland and the edge of the continental shelf west of Ireland. The marine biologists, micropaleontologists, geochemists, and sedimentologists of the US Geological Survey examined the split cores and recorded four glacial periods and two periods of volcanic ash sedimentation (Figure 12.3). Changes in the orientation of the Earth's magnetic field with depth in those cores were studied at DTM. Piggot verified the high content of radium in these sediments, but was unable to explain why it was higher than the igneous or sedimentary rocks found on land.

Figure 12.2   Cores, split to show sediment variations taken at intervals between the Grand Banks of Newfoundland and the edge of the continental shelf west of Ireland. See Figure 12.3 for locations. *Source*: As Figure 12.1, p. 215, figure 8. By permission of the American Association for the Advancement of Science.

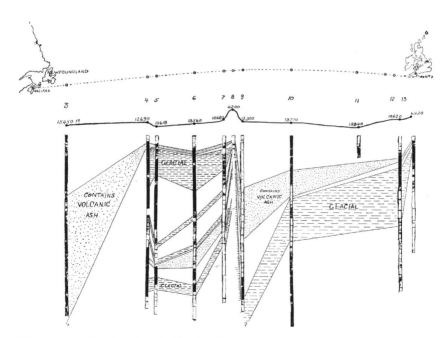

Figure 12.3   Core locations in North Atlantic and interpretation of sedimentary layers on the basis of petrographic analysis. *Source*: As Figure 12.1, p. 214, figure 6. By permission of the American Association for the Advancement of Science.

The strontium content of the pure fluids in deep-sea sediments of Creta-ceous to Quaternary age was also of concern to investigators. In the 1970s A. Hoffmann's results indicated that within present limits of error the iso-topic composition of the pure fluids is identified with that of seawater at the time of deposition.[25]

P. E. Hare measured the amino acids in a sediment core from the Cariaco Trench in the Caribbean Sea.[26] Individual amino acid concentration, as well as the amino sugar and ammonia concentrations, was listed at three depths: 1 m, 80 m, and 108 m. The carbonate phase of the sediments was characteristic of invertebrate calcified tissue, whereas other phases were suggestive of land plant origin. Hare was confident that the organic matter would eventually reveal the sources of the sedimentary organic matter, its diagenesis, and even its contamination by the migration of soluble organic matter.

With the discovery of abundant life around the hydrothermal vents on the ocean bottom, J. D. Frantz volunteered as part of the crew of the submersible ALVIN to study those deep-ocean vents at the Juan de Fuca Ridge off the coast of Washington. He developed an instrument to measure the hydrogen fugacity of high-temperature fluids. As a measure of the fluid's oxidation state, it had direct bearing on the microbial activity within the hydrothermal structures. By using a gold–palladium semi-permeable membrane, he was able for the first time to measure hydrogen fugacities at various temperatures. Needless to say, deployment of his apparatus at the active vents was at no small risk. As already mentioned in Chapter 11, the vents are considered the prime location for the origin of life on Earth.

Magnetism

The pressure effect on the critical temperature of magnetization of iron was found to be negligible up to 3.6 kbar in investigations by L. H. Adams and J. W. Green in the early 1930s.[27] They concluded that the nickel–iron core had little influence on the Earth's magnetic field because the core temperature was well above the Curie point! The Curie point was investigated for a large number of materials by E. Posnjak during 1936 and 1937, and the effects of solid solution on the magnetic properties of spinels were the subject of inves-tigation for many years to follow. The effect of pressure on the Curie point of iron was studied, but experimental difficulties did not yield acceptable results. A 6 °C rise in Curie temperature in cadmium-magnesium ferrite, a solid solution of the spinel type, was observed up to 12 kbar.[28] Twenty years later, using the same technique as Adams and Green, L. Patrick found a very small change up to 8 kbar.[29] A new inductive method was applied by Timofeev and colleagues in 1999 to obtain the magnetic susceptibility of iron up to 100 kbar.[30]

The magnetic structures of chalcopyrite, cubanite, ardennite, tourmaline, and the spinels were studied in great detail in the 1960s and 1970s. The effects of solid solution were determined in the systems $FeO-Fe_2O_3-TiO_2$ and $FeCr_2O_4-Fe_3O_4$, with special attention to the regions of antiferromagnetism and ferromagnetism.[31] Posnjak's data published in 1940 indicated that the Curie temperature of the ferrous sulfides rose appreciably with increasing sulfur content.[32] In addition to a study of magnetite in basalts, Osborn and colleagues observed the composition of magnetites in subalkaline volcanic rocks as well as those in calc-alkaline rocks.[33] An examination of the submarine basalts of the mid-Atlantic Ridge revealed that the magnetic properties could be related to the process of progressive oxidation with age, that is, distance from the ridge axis.[34]

Electrical conductivity

Inferences can be drawn about the electrical conductivity of the Earth from the analysis of transient components of the magnetic field. In addition, heat transfer by radiation is influenced by electrical conductors since certain types absorb strongly. For these reasons S. P. Clark studied the electrical conductivity of the olivines.[35] He recognized the potential of a geothermometer based on electrical conductivity. H-k. Mao, in the early 1970s, measured the electrical properties of the olivines at high pressures up to 300 kbar in a diamond cell.[36] He demonstrated that the electrical conductivity is related to the amount of iron present. In the pressure interval studied, the conductivity of fayalite increased by six orders of magnitude, whereas $Fa_{42.5}Fo_{57.5}$ increased by only three orders of magnitude. He concluded that pressure increases heat transfer by electronic and possibly excitonic processes. In 1976, these measurements were extended to 1.2 Mbar by Mao and P. M. Bell in a diamond cell.[37]

In order to interpret the low-velocity zone of the mantle, Murase and colleagues obtained the electrical conductivity in partially molten peridotite.[38] The effective bulk conductivity as a function of volume fraction of melted peridotite was obtained at 15 kbar. Their results indicate that a volume fraction of melt in the upper mantle of less than 0.03 in the range 1150–1200 °C would account for the observed conductivity. They believed, therefore, that the low-velocity region is a region of partial melting.

Aqueous solutions commonly consist of mixtures of ions and molecular species. Identification of the prevalent ions is fundamental to the problem of material transport. A substantial study, published in 1907,[39] of the electrical conductivity of aqueous solution was undertaken by A. A. Noyes at MIT with the support of CIW. Some seven chlorides, nitrates, and sulfates exhibited a decrease in conductivity with temperature. Adams and R. E. Hall later measured the electrical conductivity of dilute and concentrated NaCl solutions up to 4 kbar.[40] The conductivity went through a maximum, but they were concerned about the influence of pressure on other physical factors. By measuring the electrical conductance in aqueous solutions such as $MgCl_2$-H_2O and $CaCl_2$-H_2O, J. D. Frantz and W. L. Marshall, working in the late 1970s, observed a trend toward association of ions above 400 °C at 1 and 2 kbar.[41] Above 500 °C there was an increase in the degree of formation of neutral species (ion pairs). Mass transport would appear to be as associated species and such complexes would contain the bulk of the metals in solution.

Seismology

Of special significance was the theoretical contribution made by L. H. Adams and E. D. Williamson in 1923 when they deduced a formula that related the compressibility and density of rocks to the seismic velocities of the longitudinal and shear waves.[42] In this way, the laboratory measurement of the density and compressibility of rocks and minerals constrained the kinds and proportion of phases in those portions of the Earth where the seismic velocities were known. Observational seismology had been recommended by the Van Hise Committee as early as 1903 and proposed on numerous occasions thereafter. The Director of the Geophysical Laboratory, A. L. Day, was appointed chairman of the CIW Advisory Committee in Seismology in 1921. On his Committee's recommendation, a program of study was outlined and a Seismology Laboratory built in Pasadena, California, in 1926 in cooperation with the California Institute of Technology. In October 1930, CIW established a Department of Geophysics headed by Dr. Beno Gutenberg with its headquartered at the Seismology Laboratory. The studies were administered by the Committee until January 1, 1937 when the Seismological Laboratory was turned over to Caltech. That Laboratory was primarily concerned with natural earthquakes.

In a letter to Dr. Day dated March 14, 1923, Dr. L. H. Adams proposed the utilization of artificial earthquakes for investigating the Earth's interior. He noted that when five tons of aluminum perchlorate were detonated at Oldebroek, Holland, a seismic record was obtained at Woolwich, England. Adams also suggested that allotments of explosives could be obtained from the War Department at small cost, and that abandoned mine shafts in hard rock could no doubt be used for detonation sites. The new field of "explosion seismology" was thereby initiated. Following World War II, DTM undertook a cooperative program with the Geophysical Laboratory, with staff members J. W. Greig, J. L. England, and G. L. Davis selecting the seismometer sites. They selected the sites for their geological advantages and occupied those sites to receive signals from quarry blasts and the destruction of old military explosives. Summaries of the principal seismic data for several continental regions of the USA were presented in 1953.[43]

In order to explain the 3 percent increase in seismic velocities between 150 and 200 kbar, Clark and A. E. Ringwood suggested that it resulted

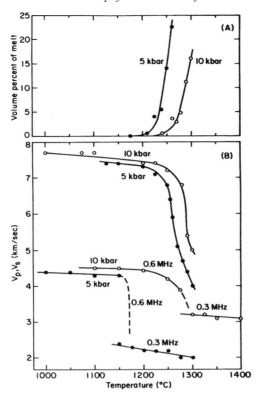

Figure 12.4   Shear and compressional wave velocities versus temperature for spinel lherzolite at 5 kbar (solid circles) and 10 kbar (open circles). *Source*: From T. Murase and H. Fukuyama, "Shear wave velocity in partially molten peridotite at high pressures," *CIW Year Book* 79 (1980), 309, figure 27.

from the transition from pyroxene pyrolite to garnet pyrolite.[44] To test this hypothesis, calculations were made of the velocities in those rock types using available data on elastic constants, densities, and measured seismic velocities in rocks and contained minerals.[45] The researchers found that the seismic velocity in garnet pyrolite is about 2.5 percent greater than that of pyroxene pyrolite. Aside from different pressures and temperatures, they concluded the difference was insignificant and that the change in velocity at that depth could be due to the transition from pyroxene pyrolite to garnet pyrolite.

Direct measurements of the elastic wave velocities in partially molten mantle minerals were needed to understand the nature of the low-velocity zone in the earth. Toward this end, T. Murase and I. Kushiro measured the velocities in a spinel lherzolite as a function of temperature[46] using a modified pulse transmission technique in a gas-media, high-pressure apparatus.[47] The results at 5 and 10 kbar (Figure 12.4) show that 20 °C above the beginning of melting the velocity of the compressional wave drops sharply.[48] Special techniques were applied to measure the degree of melting and the distribution of melt among the grains. A mere 5 percent decrease in the compressional velocity, as observed for the low-velocity layer of the upper mantle, implies a maximum of 7–9 volume percent melting. The researchers also studied the effect of melting on the shear-wave velocity. As apparent in Figure 12.4, above 1150 °C the shear wave disappeared even though no melt was observed there at 5 kbar. At 10 kbar, however, the shear wave decreased with the onset of melting and disappeared above 1300 °C. The results clearly support the predictions of D. Z. Anderson and H. Spetzler that the shear wave would be more sensitive to melting than the compressional wave.[49]

13

EXTRATERRESTRIAL PETROLOGY

Meteorites

In 1905 T. C. Chamberlin reported on his progress on Grant No. 115 (a continuation of Grant No. 31), on one of the fundamental problems of geology.[1] He developed in more definitive terms several stages of the meteoritic hypotheses of the Earth's origin as proposed by J. N. Lockyer in 1887[2] and G. H. Darwin in 1889.[3] The nickel–iron core of the Earth was at one time believed to be surrounded by a zone of mixed iron and silicate called pallasite. Because of the supposed similarity between some meteorites and the interior of the Earth, the pallasites were studied both for their structure and range of composition by L. H. Adams and H. S. Washington, who published their findings in 1924.[4] They conceived of the Earth not as a huge meteorite, but "a body similar to those of which meteorites are but fragments representing different parts of the whole mass. On disruption, therefore, the Earth would yield all of the known kinds of meteorites."[5]

Meteorites also contribute to an understanding of the origin and history of the solar system. For this reason, a number of studies were embarked upon from the 1960s onwards. S. P. Clark and G. Kullerud undertook a study of Fe-Ni-S and Fe-Ni-P to establish a buffered system of taenite and kamacite with troilite or schreibersite in order to derive the temperature of formation of the meteorites.[6] While in residence, P. Ramdohr examined with Kullerud over 340 polished sections of stony meteorites to characterize in a systematic way the mineralogy of the opaque phases.[7] In typical fashion, Ramdohr discovered many new minerals as well as previously described minerals hitherto not previously observed in meteorites. The Fe-Cr-S system was studied by A. El Goresy and Kullerud to account for the Cr-bearing compounds in meteorites.[8] They found the sulfides responded more readily to shock impact than silicates, thereby explaining their disequilibrium relations. The relationships of chromiferous diopside, chromite, and kosmochlor were studied by H. S. Yoder and Kullerud who found that pressure was an important variable in determining the assemblages in meteorites versus terrestrial rocks.[9]

A special effort was made to determine the ratio of silicates to opaque minerals in the chondrites, which comprise 85 percent of all known meteorite falls, with quantitative microprobe analyses.[10] The Ashmore meteorite

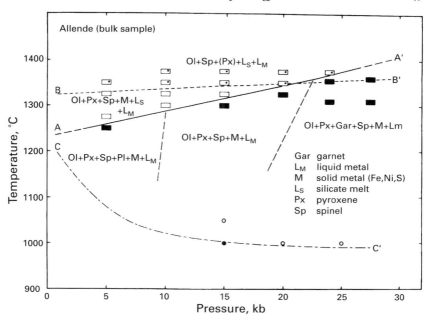

Figure 13.1 Melting relations of the Allende carbonaceous chondrite. Solid curve A–A' is solidus of the silicate phases and dashed curve B–B' is a liquidus curve of the metallic phases under anhydrous conditions. The dashed curve C–C' is the solidus of the silicate phases under hydrous conditions. *Source*: From I. Kushiro and M. G. Seitz, "Experimental studies on the Allende chondrite and the early evolution of terrestrial planets," *CIW Year Book* 73 (1974), 450, figure 159.

(Texas), for example, contained 87.2 percent nonopaques and 12.8 percent opaques by volume.

The oxidation state of iron in meteorites was of special interest to D. Virgo.[11] From the ^{57}Fe Mössbauer spectra of an untreated Orguerl meteorite, he found the principal Fe-bearing minerals to be magnetite and a layered silicate, possibly a Fe-rich serpentine with dominantly trivalent iron. Studies of the nuclear tracks in whitlockite, β-$Ca_3(PO_4)_2$, produced by high-energy proton bombardments, were undertaken by M. G. Seitz in the Saint Severin meteorite.[12] His goal was to obtain some measure of the cosmic-ray exposure ages. Seitz joined with D. S. Burnett and P. M. Bell to determine the chemical balance of thorium, uranium, and plutonium for the same meteorite and concluded that 96 percent of the nucleosynthesis took place about 8 billion years ago.[13] It now appears possible to date processes involved in the early evolution of the solar system.

Over the years the Allende carbonaceous chondrite has received special attention from Laboratory staff because of its potential applications to the early evolution of planets of alleged chondritic composition. The *P–T* diagram (Figure 13.1) for the Allende meteorite up to 30 kbar with and without H_2O was investigated by I Kushiro and Seitz in the 1970s.[14] They demonstrated that separation of the phases would yield the layered structure presumed for the Earth. Other experiments dealt with the partitioning of elements between the metal, oxide and silicate portions.

Although boron is rare in meteorites, it does have the potential for recording the pressure and thermal history of the rock. The partitioning of boron was achieved by Seitz through an etching technique that revealed the spallation record tracks in whitlockite.[15] The chondrules of the Allende chondrite were expected to be depleted of boron because of their assumed condensation from high temperatures, but instead contained high concentration of boron. The result remains a mystery in the light of the volatility of boron. The fassaite of the Allende meteorite was also interesting because it was found to be iron-free and contained trivalent titanium, confirmed by high-resolution optical spectra.[16] The chemical incompatibility of its oxidation state with other minerals in the meteorite, such as metallic iron and andradite, was evident.

The amino acids were identified by J. R. Cronin who investigated the process whereby abiotic organic compounds could be formed from the precursor compounds in carbonaceous chondrites.[17] These results support the view that the origin of life is preceded by a period of abiotic synthesis and accumulation of the chemical components of the first organisms.

In 2002 G. D. Cody and colleagues devised a series of eight independent solid state Nuclear Magnetic Resonance (NMR) spectroscopy experiments designed to establish a self consistent picture of the chemical structure of organic solids entrained in the Murchison carbonaceous chondrite.[18] The organic matter carries within its macromolecular structure a history spanning its formation in the cold dense molecular clouds, through cloud collapse and the formation of the early solar nebula, to aqueous processing in the meteorite's asteroidal parent body. Using these NMR methods, the investigators were able to prove that, contrary to some popular opinion, there exist no large hydrogen-poor polycyclic domains. The only hydrogen-poor carbon-rich domains detected were nano-diamonds, possibly of interstellar origin. The extraterrestrial macromolecular structure consists of small highly substituted aromatic molecules linked by short highly branched aliphatic chains. These results provided for the first time a fully constrained analysis of the bulk of the organic matter contained within carbonaceous chondrites.

Meteorites frequently contain minerals formed at high pressures through shock. For this reason they are important in understanding the phase changes that take place at depth in the Earth's mantle. With the discovery of ringwoodite, the spinel form of olivine, N. Boctor and colleagues were able to suggest that the pressure was between 100 and 225 kbar during impact

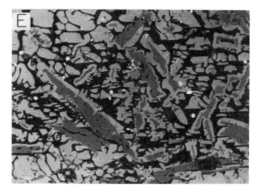

Figure 13.2 Armalcolite quench crystals (dark gray) with ilmenite reaction rims (gray) and metal blebs (white). *Source*: From H.-K. Mao, A. El Goresy and P. M. Bell, "Evidence of extensive chemical reduction on lunar regolith samples from the Apollo 17 site," *CIW Year Book* 73 (1974), 471, figure 173.

metamorphism.[19] The presence of majorite in association with ringwoodite suggested to them that a large pressure gradient existed in the order of 100 kbar to >300 kbar. Other features such as veins of inhomogeneous glass from incipient melting, fracturing, undulatory extinction, and mosaicity are also indicative of high shock pressures.

Lunar samples

The return of the successful Apollo missions with samples of the Moon beginning in 1969 provided one of the most exciting scientific opportunities for the following decade. The incredibly fresh and unaltered character of the rocks greatly facilitated their study, but the fine grain and shock metamorphism of the minerals were the principal challenges. The characterization of the fine-grained material generated a new array of techniques and the unique conditions of rock formation on the Moon led to the new field of comparative petrology. Needless to say, the daily excitement of discovery has not been equaled by the arrival of any other set of specimens. The initial stages of inquiry were predominantly detailed mineralogical studies, followed by experimental studies of both the natural samples and synthetic analogues, and then the testing of various models of the Moon's composition and structure. The entire array of sample types (rocks, breccias, glass fragments, and soils) was investigated by the staff members.

Because of the special skills of the staff, the opaque minerals received detailed attention. Ilmenite was the major opaque phase, but members of the chromite-ulvöspinel series, the newly discovered armalcolite series (Figure 13.2),[20] as well as troilite and metallic iron alloy were studied.

(Armalcolite was also discovered independently by several laboratories, and subsequently named by combining the initial letters of the names of the astronauts *Arm*strong, *Al*drin, and *Co*llins.) S. E. Haggerty and H. Meyer found that the spinels were bimodal at some sites, but other samples exhibited a complete series of solid solutions.[21] A new pyroxenoid was discovered by D. H. Lindsley and C. W. Burnham,[22] which was then prepared synthetically at high pressures. The new phase, pyroxferroite, is apparently metastable and had persisted in that state for at least 3 billion years. The olivines also appeared to be bimodal in some samples. The Cr content of olivine was more than twice that of Earth olivines, and more importantly, was in a reduced state.[23] The first demonstration of Fe-Mg ordering in any olivine was made by L. W. Finger on the lunar material.[24]

The pyroxenes were studied in exceptional detail by F. R. Boyd whose electron microprobe data clearly reflected their chaotic crystallization behavior.[25] The zoning of the pigeonites and oscillatory augite rims, for example, suggested cooling and mixing with new magma batches. Two distinct rate-determining steps were found by Virgo in the cooling of two lunar pigeonites.[26] A steady state of Fe-Mg ordering in pigeonite was achieved at about 810 °C requiring a few hours, then an exceptionally slow rate to about 480 °C below which no further annealing was possible. (The ^{57}Fe Mössbauer resonance spectroscopy technique for determining valence state and geometrical configuration of iron in a crystal structure had been previously proved useful in kinetic studies of terrestrial pyroxenes.) The plagioclase contained Fe^{2+} and its site preference was determined. The range of shock features in the plagioclase were particularly impressive.

The lunar glasses were studied with the new high-resolution, optical-spectra apparatus of Mao and Bell.[27] Some of the glasses were igneous in origin and others had formed from the meteorite impacts. Various colored glasses were found to result from the reduced states of Fe and Ti. A particularly exciting event, although short lived, was the discovery of "rust" (goethite and akaganéite) on some of the lunar specimens,[28] eventually attributed to accidental contamination in the Earth's atmosphere. The event was not only an indication of the great care taken by observers, but also of the advanced state of the art of characterizing fine materials.

Evidence of extensive chemical reduction in the lunar regolith was uncovered by Mao and his colleagues, whose data led to the view that hydrogen-rich gas, implanted in the upper portions of the regolith by the solar wind, was released when melted rock came in contact.[29] A quenched outer skin of lava probably tended to confine the gas. Large-scale chemical reduction was displayed by the mineral assemblages and the bubbles of trapped gas, particularly in the orange glass. The extent to which this process erased the original oxidation state is not known.

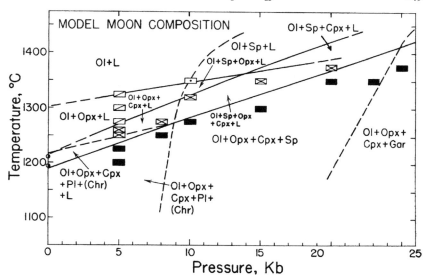

Figure 13.3 Phase assemblages obtained experimentally on a model moon composition. (Ol, olivine; Opx, orthopyroxene; Cpx, clinopyroxene; Gar, garnet; Sp, spinel; Chr, chromite; Pl, plagioclase; L, liquid.) *Source*: From I. Kushiro and F. N. Hodges, "Differentiation of the model moon," *CIW Year Book* 73 (1974), 455, figure 162.

The experimentalists also contributed to the understanding of the formation on the Moon of the principal rock types, called basalts and anorthosites, even though they were quite different in composition and mineralogy from their Earth-bound namesakes. Because of the extreme rarity of the lunar samples, A. Muan and F. Schairer made a synthetic analogue and studied its behavior at a series of temperatures in iron crucibles.[30] This "basalt" composition yielded pyroxene on the liquidus at 1185 °C and had a solidus between 1075 °C and 1090 °C with plagioclase and ilmenite. These values are not greatly different from those of some Earth basalts. Later, as more material became available, lunar samples themselves were studied at a range of pressure and temperature (Figure 13.3) by F. N. Hodges and Kushiro.[31] Of the three models of lunar composition they tested, the model composition of R. Ganapathy and E. Anders appeared to fit the observations best.[32] A most interesting observation by Bell and Mao in support of the high-pressure experiments was recognition that a spinel + two-pyroxene symplectite in olivine had the bulk composition of garnet.[33] The reaction of garnet and olivine to the symplectite assemblage had been previously demonstrated to be a high-pressure reaction. Another set of pioneering experiments was performed by Seitz, who was concerned about the chemical fractionation that resulted from the volatilization of materials during the impact events.[34] He showed in a vacuum furnace that all of the alkalies and some of the iron were lost by volatilization in short heating events. These experiments were precursory to the major program of study, described below, most pertinent to the origin of the solar system.

Condensation petrology

An understanding of the evolutionary processes of the solar system requires data on pressure, temperature, and composition at various times and places as the solar nebula collapses. The fundamental issue is whether the minerals formed by direct condensation from a gas by partial evaporation of precursor phases or by crystallization from an intermediate liquid phase of the proto-solar system. Experiments conducted at the Geophysical Laboratory by B. O. Mysen, D. Virgo, and I. Kushiro bear on these processes. The materials found in meteorites are believed to be representative of the oldest solar system material. For this reason, the investigators worked on the very low-pressure region of stability for minerals such as akermanite, diopside, corundum, spinel, and hibonite, found in the carbonaceous chondrites.[35] By means of a Knudsen-cell technique, in a high-vacuum, high-temperature furnace, they established P–T curves separating the crystal, vapor, and liquid regions. From the experimental results, they suggested that for the solar-gas composition, the pressure would have to exceed 10^{-2} bar for liquid to form, increasing with decreasing oxygen fugacity. In general, it appears that the early solar nebula resulted from gas–crystal reactions in the absence of melting at pressures below 10^{-4} bar and at an $f(O_2)$ at least three orders of magnitude below that of the iron-wüstite buffer. In short, these dramatic experiments place severe constraints on the collapse of the solar nebula and emphasize the systematic chemical differences between the terrestrial planets as a function of their distance from the Sun.

Other constraints are placed on the gaseous planets by experiments in a totally different realm. Materials that are normally gaseous condense to form liquids and crystals at high pressure. Studies up to 5.5 Mbar are particularly pertinent to the early evolution of the solar system as well as the interior of Jupiter, for example. The P–V curves for crystalline hydrogen, deuterium, argon, neon, xenon, and oxygen have been determined, and, in some cases, their crystalline structure determined with synchrotron-generated radiation.[36] The first single-crystal structure determination of n-H_2 at 54 kbar with conventional X-ray diffraction[37] was of fundamental interest to condensed-matter and planetary physicists. In addition, methane and water, important in the Giant Planets, were also studied. In this way Mao, R. J. Hemley, and a large number of colleagues were able to set limits on the conditions required to collapse these gases in the nebula.

14

ASTROBIOLOGY

The discovery of abundant life at the ocean bottom near a smoker with elevated temperatures and an array of gases, initiated an entirely new view of the origin of life on Earth. The possibility arose that organic reactions essential for life might be enhanced by the high temperatures and high water pressures at the ocean bottom. This idea, championed by biologist Harold Morowitz at George Mason University, led to a joint experimental research program by Morowitz and Robert Hazen, who also taught at George Mason University. Hazen enlisted his colleagues George Cody, Marilyn Fogel, and Hatten Yoder in performing experiments on hydrothermal organic synthesis to provide insight into the origin of life. The first reactions investigated in 1996 were focused on the reductive citrate cycle, a primary biological pathway for carbon fixation. An attempt was made to reverse-engineer the primitive cycle by observing the pathways by which citric acid is decomposed. The pathway to propene and carbon dioxide was selected as one that might have great promise for generating a primitive reductive-type cycle.

The new line of research was given a major boost when Dr. Wesley T. Huntress, Jr. was made Director of the Geophysical Laboratory in the fall of 1998. He was one of the early developers of the science of astrochemistry and founded NASA's program in astrobiology. The Geophysical Laboratory and the Department of Terrestrial Magnetism together received an award for research in astrobiology at the same time Dr. Huntress arrived on campus. The objective was to provide a suite of models for the atmosphere and surface chemical environments during the Earth's formative stages. The chemical processes were to be outlined that would lead to pre-biotic organic chemistry, directed by mineral catalysis and template formation, that could be inherited by a subsequent biochemistry in the transition from a chemically-driven environment to one driven biologically.

Interstellar medium

The chemical evolution of the interstellar medium is currently being investigated by Dr. Huntress and his former colleagues at the Jet Propulsion Laboratory. Studies of the gas-phase, ion-molecule reactions in combination with astronomical observations determine how the chemical evolution proceeds.[1]

The model starts with atomic matter in the "empty" space between stars to molecular matter as the medium condenses first to UV-transparent clouds where distonic molecules are formed to denser shielded regions where cosmic ray ionization leads to the formation of complex organic molecules. The goal is to trace chemical evolution through the entire timeline from the interstellar medium to the organic material in a young stellar proto-planetary disk. In that way they hope to identify the starting material available for sparking the origin of life. The compositional data from the analysis of cometary material will be used to help determine the chemical relationships between comets in the Solar System and the original interstellar molecular cloud out of which the proto-stella nebula condensed. They suspect the chemical evolution proceeded in distinct steps controlled by kinetics, and chemical cycles developed with the evolution being punctuated by a series of emergent states. Identification of such cycles will be important to recognizing the precursors to biochemical cycles, such as the citric acid cycle, which is vital to the emergence of biology itself on a young planet.

Hydrothermal pathway

The goal was to demonstrate that a hydrothermal environment would provide a favorable pathway into the reductive citrate cycle.[2] Homocatalytic reactions observed in citric acid-H_2O experiments encompassed many of the reactions found in modern metabolic systems, i.e. hydration–dehydration, retro-Aldol, decarboxylation, hydrogenation, and isomerization reactions. These were demonstrated to be feasible nonenzymatically and under realistic conditions of concentration, pressure, and temperature. Entry with the reductive citrate cycle (Figure 14.1) was achieved via the synthesis of citric acid from the simple molecule propene. With this knowledge, it became evident that entry into the cycle could be achieved at several points. In short, the options for initiating sources of energy for the earliest life forms became multiple. At the earliest times, perhaps 4 billion years ago, life may have begun, possibly obliterated by asteroid bombardment, but then started up again at the same or a new point in the cycle. The deep submarine hydrothermal vent systems were, then, ideal environments for the emergence of the earliest life. The traditional view of life's origin on Earth was focused on processes near the photic zone at the ocean–atmosphere interface where ionizing radiation provided the energy for prebiotic organic synthesis. The hydrothermal experiments clearly illustrate an alternative, particularly at the earliest stages, for life initiation in deep hydrothermal zones. Nevertheless, the investigators caution that these primitive reactions may be far removed from the archaic carbon-fixation chemistry.[3] Life emerged because the geochemical conditions on Earth were both kinetically and thermodynamically favorable for the primitive biochemistry.

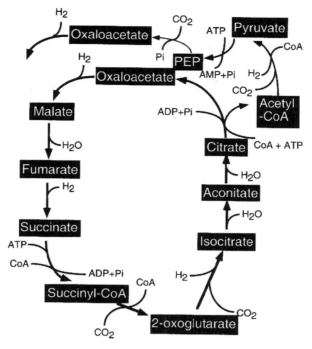

Figure 14.1 Map of reductive citrite cycle (RCC). The cycle is as follows: Beginning with acetyl CoA, carbon insertion leads to the formation of pyruvate. Phosphorylation of pyruvate at the cost of one mol of ATP (adenosine triphosphate) leads to the formation of phosphoenol pyruvate (PEP). Electrophilic addition of CO_2 leads to the formation of oxaloacetate. Reduction of oxaloacetate forms malate. Dehydration of malate forms fumarate. That is then reduced to form succinate. Succinate is activated to succinyl CoA at the cost of one mol of ATP. Carbon insertion leads to the formation of 2-oxogluterate. A second carbon insertion leads to the formation of isocitrate. Isocitrate is isomerized through aconitate to citrate. The splitting of citrate to form one mol of acetyl CoA and oxaloacetate completes the cycle. This final step reveals the autocatylic nature of the network, in that the cycle doubles the number of CO_2 acceptors with each revolution. From G. D. Cody, N. Z. Boctor, R. M. Hazen, *et al.*, "Geochemical roots of autotrophic carbon fixation," *Geochim. Cosmochim. Acta* 65 (2001), 3558, figure 1. Reprinted with permission of Elsevier.

Mineral catalysis

The special role of transition metal sulfides in the origin of life was proposed by G. Wächterhäuser in 1988 wherein iron sulfide, for example, acts as both a catalyst and a source of energy for the earliest life.[4] Through the reaction of alkyl thiols and carbon monoxide with iron sulfide, Cody and colleagues generated a suite of organometallic compounds that were characterized with ultraviolet–visible and Raman spectroscopy at elevated temperatures (200–300 °C) and pressures (1–2 kbar).[5] In a following detailed study, a wide array of metal sulfides at range of temperatures and pressures were examined.[6] First, they demonstrated that the reactions are surface catalyzed. The sulfides FeS and $Fe_{(1-x)}$ S are unique in that they show some evidence of dissolution thereby complicating the interpretation of heterogeneous vs. homogeneous catalysis. With the exception of CuS, each of the metal sulfides promotes reactions that mimic key intermediate steps manifest in the autotrophic enzyme, acetyl-CoA synthase. The most productive catalysts were NiS, Ni_3S_2, and CoS in terms of the yield of decanoic acid. From these results, it would appear that mineral catalysts were undoubtedly required in a pre-enzymatic world to initiate a broad range of organosynthetic reactions toward producing the first biomolecules. Needless to say, there is an abundance of metal sulfides in close proximity to the hydrothermal hot springs at the ocean bottom.

Chirality

One of the most distinctive biochemical signatures is the selectivity for chiral molecular species, notably L-amino acids and D-sugars. Most synthesis reactions yield essentially equal amounts of L- and D-enantiomers. The production of the observed chiral in living organisms could not be accounted for until Hazen and colleagues suggested that common minerals, such as quartz and calcite, provided crystal faces that were chirally selective.[7] Such minerals possess pairs of crystal faces whose surface structures are mirror images of each other, and are ideally suited to select and concentrate L- and D-centers in molecules such as amino acids. Admittedly, the crystals have approximately equal numbers of left- and right-handed surfaces, but the origin of life is probably a local event and once initiated could become dominant. Experiments were therefore conducted to demonstrate the selection process. When calcite was immersed in a racemic aspartic acid solution, significant adsorption and chiral selectivity on pairs of mirror-related growth surfaces was observed. Those surfaces with terraced surface textures had greater selective adsorption. Large natural crystals with enantiometric faces occur commonly also as gypsum, barite, and apatite, as well as all the major-rock-forming minerals in the most common rock basalt. The investigators concluded that selective adsorption on crystal surfaces is a plausible mechanism for chiral selection and subsequent homochiral polymerization of amino acids on the prebiotic Earth.

Amino acid synthesis

A primary requirement for the origin of life is the prebiotic synthesis of amino acids. Experiments for the synthesis of amino acid were attempted in the past by (1) electrical discharges;[8] (2) ultraviolet irradiation of solutions;[9] polymerization reactions under aquifers conditions;[10] or by Fischer–Tropsch chemistry.[11] Although some success was achieved, none of those approaches resulted in the synthesis of all twenty amino acids commonly employed by organisms. In fact, some of the approaches involved conditions inhospitable for life. The most successful synthesis pathways involve the reaction and polymerizations of aldehydes and cyanides, e.g. the Strecker synthesis. An alternative to the Strecker synthesis was investigated by Brandes and colleagues in 2000 in which ammonia is reacted with pyruvic acid to yield alanine + acetic acid + carbonic acid.[12] The synthesis of alanine was rapid, reaching peak values at 100 °C within 6–24 hours at one atmosphere. The reaction is very sensitive to pH with the highest conversions at pH = 0.6–0.8. Alanine yields increased slightly with dilution until an acid : water ratio of 1 : 100 was reached at which point a sharp decline in yields was noted. The synthesis of ammonia in hydrothermal systems has been demonstrated,[13] so the alternative reaction has considerable potential as a biochemical pathway for the formation of amino acids at the earliest stages.

Protein chip analysis

A method is being developed for detecting life in a range of environments both here on Earth and on the surfaces of planets. A chip-based molecular recognition technology has been perfected to capture selectively femtomole amounts of complex mixtures of organic molecules.[14] The surface-enhanced chips spatially orient molecules that are bound to them, such that they are more efficiently uncoupled and ionized for mass spectral analysis. By selecting a series of chip surface chemistries, complex mixtures can be pulled apart. In a test on a microorganism, some sixty-two peptides or proteins were identified. Current efforts are being directed toward identifying unambiguous signatures of biological molecules. An instrument for detecting early life on other planets, for example Mars, may be in hand.

Chondrites

Initial interest in the chondrites stemmed from the hypothesis that the overall composition of the Earth's mantle is the same as the silicate fraction of chondritic meteorites.[15] Attention turned to the carbonaceous chondrites as the potential source for the introduction of life on Earth. If they indeed record prebiotic organic synthesis in the interstellar medium, then their molecular structures may be important to the origin of the terrestrial organic structures. Organic material in chondrites can be divided into two fractions: a soluble fraction composed of a complex mix of compounds, including many amino acids, and an insoluble fraction composed of a largely uncharacterized macromolecular material. With nuclear magnetic resonance (NMR) spectroscopy, Cody and colleagues examined a solvent-extracted and demineralized sample of the Murchison meteorite.[16] What emerged from the analysis was a picture of a structurally complex, extensively cross-linked, and highly aromatic macromolecule. It does provide boundary conditions for the insoluble organic residue, but the investigators note that more information must be obtained on the types and abundance of the oxygen-, nitrogen-, and sulfur-containing functional groups before the organic macromolecules in the Murchison meteorite can be defined. Only then can consideration be given to the pathway of its solar or interstellar origin.

Water in meteorites

Water is considered essential for life on Earth, and, therefore, is used as a test for life on other planets. For this reason Boctor and colleagues have analyzed nine Martian meteorites by secondary ion mass spectrometry to determine the hydrogen isotopic compositions and water contents of the mineral phases and glasses.[17] All of the crystalline and glass phases contained deuterium-enriched water of extraterrestrial origin. The water content is highly variable depending on the phase analyzed, ranging up to 1.0 wt. percent in carbonates, for example. The investigators believed, however, that the isotopic composition was modified by (1) interaction with a fractionated exchangeable water reservoir on Mars, (2) hydrogen devolatilization by impact melting, and (3) terrestrial contamination. Nevertheless, they interpreted the low δD of magmatic glass in melt inclusions to signify that the initial H isotopic signature of Mars may be similar to that of Earth.

15

MINERAL PHYSICS

The first Mineral Physics Conference was organized by R. M. Hazen and C. T. Prewitt in 1975 and sponsored by the Carnegie Institute of Washington and the National Science Foundation. In the same year the first issue of a journal, *Physics and Chemistry of Minerals*, was published by its founders and original editors S. S. Hafner, A. S. Marfunin, and C. T. Prewitt. The field of Mineral Physics was formally recognized in 1983 when the president of the American Geophysical Union, Dr. James Van Allen, appointed a committee to coordinate the subdisciplines dealing with the relationships of the physical and chemical properties of minerals to their geological behavior at especially high pressures and temperatures. The Mineral Physics Committee met in 1985 and a report was prepared by W. A. Bassett defining its "roots in mineralogy, physics, and chemistry and its boundaries with geophysics, crystallography, petrology, and geochemistry."[1] Members of the Committee were Dr. H-k. Mao of the Geophysical Laboratory and C. T. Prewitt (who later became director of the Laboratory in 1986). The goal was to correlate observed properties with inaccessible properties at depth in the Earth and other planets, i.e. composition and temperature. For example, F. Birch, some thirty years earlier, had interpreted the seismic profiles in terms of composition and phase transformation with depth in the Earth.[2] A workshop in 1986 chaired by C. T. Prewitt prepared a set of recommendations[3] for the development and use of four specific technologies: (1) synchrotron radiation facilities, (2) high-pressure, high-temperature, large-volume experimentation, (3) in-situ analytical instrumentation, (4) accelerator mass spectrometry. The report was very influential in generating financial support for Mineral Physics from the funding agencies. Since then, a wide range of techniques has been applied to deduce the properties of materials under extreme conditions.

Mineral Physics at the Laboratory evolved mainly from initial efforts in mineral optics, arc-Raman, and X-ray crystallography, followed in the 1970s by Mössbauer, laser-Raman and infrared spectroscopy, and in the 1980s by the entire array of spectral tools of modern-day physics and chemistry.

Optical

An early pioneer in the USA in the application of microscopy to mineralogical and petrological problems was F. E. Wright. His design of a petrographic microscope was adopted by a leading manufacturer and his improvements substantially advanced its use. His book, *The Methods of Petrographic Microscope Research* (1911),[4] had great influence in promoting the quantitative measurement of the optical properties of crystals.

Another major contributor to crystal optics was H. E. Merwin. He developed with E. S. Larsen special immersion media of unusually high refraction using mixtures of amorphous sulfur and selenium.[5] His dispersion method for measuring refractive indices of grains in immersion liquids is still widely used. Because of his demonstration of the relationship of index of refraction, density, and composition of glasses, he contributed to many of the phase equilibria studies of the Laboratory.

Optical absorption spectroscopy was particularly useful in the study of the nature of Fe in silicate perovskite. Some of the Fe was found to substitute for Si in the octahedral site and some of the Si appears to be in the distributed 8–12 coordinated site. Although the bands in the visible region were not well resolved, a near-infrared band was clearly resolved. Consistent with X-ray data, G. Shen and colleagues in the 1990s were able to assign most of the Fe^{2+} to the 8–12 coordinated site in perovskite.[6] Furthermore, a team of investigators using gamma-ray, resonance-absorption Mössbauer data found that the small amount of Fe^{3+} in the synthetic sample was ordered in a single octahedral site.[7] Optical absorption spectroscopy of the lower mantle minerals can provide useful information on the electronic properties of iron in lower mantle minerals and its partitioning between those minerals.

Some of the earliest studies of the Raman effect, discovered in 1928, were carried out in the USA by J. H. Hibben.[8] He provided detailed treatises on inorganic compounds in 1933 and on organic compounds in 1939. His principal successes were in the speciation of organic compounds, and he made a special effort to apply the technique to the petroleum industry. The low intensity of the Raman effect using an arc lamp and photographic plates was eventually enhanced with the advent of laser light sources and photoelectric cell recording of the spectra. Applications to the common rock-forming minerals and their melts did not begin until the arrival of S. K. Sharma as a postdoctoral fellow in 1977. Sharma installed a modern Raman spectrometer and a variety of laser sources.[9] Another Raman system fitted with a microscope was added, and a multi-channel detector was introduced in the mid-1980s.[10] These refinements led to the techniques of single-crystal micro-Raman spectroscopy as well as ultra high-pressure optical spectroscopy.

High-pressure and high-temperature Raman spectroscopy have been used to demonstrate the stability of magnesite ($MgCO_3$) at mantle pressures and temperatures, which is important in determining the carbon budget deep in the Earth.[11] The high-pressure form of carbon (diamond) also provides an important source of carbon in the mantle. An investigation in the early 1990s carried out micro-Raman measurements on diamond anvils up to 3 Mbar and found some evidence for structural transformations under the large non-hydrostatic stresses of the anvils.[12] A displacive phase transition associated with pressure-induced mode softening has been identified by Raman spectroscopy on stishovite, which undergoes a transition to a $CaCl_2$ structure at ~500 kbar.[13] Because of the dominant role of silicate perovskite, $(Mg,Fe)SiO_3$, in the Earth's lower mantle, various vibrational spectroscopic investigations in the infrared and Raman range have been undertaken.[14]

Compressibility

The structural response of minerals to pressure is of fundamental importance to earth sciences. Knowledge of the differential response of the metal-oxygen bonds in solids has proved valuable in predicting the physical properties and stabilities of the minerals. In 1977 Hazen and Prewitt developed an empirical relationship between bond lengths and compression.[15] In addition, they found a simple linear relationship between compression and thermal expansion. Because garnets are particularly important in the mantle, Hazen and L. W. Finger measured the compressibility of pyrope and grossular.[16] They found that larger polyhedra with cations of lower charge are more compressible than smaller polyhedra with cations of higher charge, and that distorted polyhedra tend to become more regular at high pressure. The micas that have extremely anisotropic bonding characteristics responded as expected: the bulk of the volume change (75 percent) took place in the interlayer volume, whereas the intralayer compression was similar to that in garnets.[17] The rigid structural units in zircon, $ZrSiO_4$, on the other hand, were found to be less compressible than any other measured compound with tetrahedrally coordinated silicon.[18] Open framework silicates such as sodalite and scapolite compressed principally by deformation of the Al-Si tetrahedral framework coupled with Na-O and Ca-O bond compression, respectively.[19]

Omphacite is important at depth in the Earth especially in the eclogites. It was found to be significantly less compressible than diopside,[20] which is vacancy-rich. The three polymorphs of Al_2SiO_5 were studied to evaluate how the bonding influenced compressibility in the same bulk composition. The Al-O bond compression seems to control the change as the octahedral compression decreases from kyanite[21] to sillimanite[22] to andalusite.[23] The comparative compressibility of the end-member feldspars, major components of igneous and metamorphic rocks, was also determined.[24] Low-albite and high-sanidine show almost identical behavior on compression.[25] This implies that the aluminum-silicon order within the feldspar framework does not have a major effect. Anorthite is less compressible than either albite or sanidine and is attributable to the greater charge at the interstitial cation. A phase transition in anorthite at a pressure between 25.5 and 29.5 kbar is related to a further ordering of the calcium on two of the four sites within the structure.

The measure of compressibility is particularly important relative to the thermal expansion, both of which are a part of the equation of state of a mineral.

Electrical

Conductivity is a critical parameter in modeling the temperature, composition, and magnetism in the Earth. With the electrical conductivity, D. C. Tozer considered it possible to estimate the temperature distribution in the Earth's mantle.[26] Unfortunately, subsequent heat flow measurements resulted in the conclusion that some regions of high electrical conductivity were due to the presence of fluids and not high temperatures (S. Sacks, personal communication, 2003), so care must be taken in interpreting electrical conductivity data in terms of other parameters. Mao and P. M. Bell measured the electrical conductivity of fayalite and spinel up to 300 kbar and found an experimental increase of six orders of magnitude.[27] They believed that a new mechanism of conduction employing an efficient charge-transfer process was induced by high pressure.

With the ability to pressurize materials in the megabar range, M. I. Eremets and colleagues produced solid xenon and found the resistance changed from semiconducting to metallic at pressures between 1.21 and 1.38 Mbar.[28] A rare-gas solid had in fact been shown to be a metal by the electrical measurements from 300 K to 27 m K.

The discovery of superconductivity was made in 1911 by H. Kamerlingh-Onnes in Hg at 4 K.[29] The crystallography of the cuprate high-temperature superconductors was extensively studied by scientists at the Geophysical Laboratory.[30] J. H. Eggert and H-k. Mao *et al.* contributed to the discovery of superconductivity in one of these compounds at temperatures as high as 164 K.[31] Nevertheless, the high temperatures in the Earth suggest that it is unlikely that superconductivity will affect the interpretation of the conditions at depth.

Elasticity

Interpretation of the properties of the Earth from seismology requires knowledge of the elastic behavior of the potential constituent minerals. Seismic evidence that the Earth's inner core is elastically anisotropic was a major concept amenable to analysis from first principles. Investigators assumed that the core was pure iron and calculated the elastic constants.[32] With a simple texture of hexagonal close-packed crystals they obtained an encouraging level of agreement with the seismological observations. Experimental data up to 2.2 Mbar indicated that the simple texture involving crystal alignment was needed to explain the anisotropy.[33] They found strong lattice strain anisotropy in the iron samples, and also suggested that a preferred slip system may be involved.[34] By considering the axial ratio of hexagonal close-packed iron at the high temperatures of the core, it was thought a model of polycrystalline texture in the inner core with basal planes partially aligned with the rotation axis would account for the inner-core anisotropy.[35]

Another test was made with elastic parameters of the peridotite-based, upper-mantle model. The acoustic velocity contrast of the olivine-wadsleyite transition in $(Mg,Fe)_2SiO_4$ was used to constrain the olivine content of the mantle.[36] With a full set of elastic moduli measured at pressures between 30 and 160 kbar with Brillouin scattering in a diamond anvil cell, the investigators found that the mantle probably contained about 40 percent olivine. That value is clearly below the 60 percent olivine in peridotites. After considering the effects of solid solution, presence of other phases, and texture, they concluded the olivine content of the mantle could be larger if the pressure derivatives were lower than their measurements, and further study was needed. Accordingly, new data were obtained on the elastic moduli of forsterite and the researchers came to the same conclusion that the olivine content of the mantle was significantly below that in the peridotite model.[37] Again, more data were needed on the wadsleyite phase in the same pressure range before affirming their previous conclusion. Those data were obtained and the new results were found to be consistent with an olivine fraction in the upper mantle of 30–50 percent.[38]

Equation of state

A reliable equation of state is essential for computing the density profiles along estimated geotherms and deducing the (Mg+Fe)/Si ratio in the lower mantle over the entire range of pressure–temperature conditions. Toward this end, investigators obtained synchrotron X-ray diffraction data on magnesiowüstite $(Mg_{0.6}Fe_{0.4})O$ to 300 kbar and 800 K.[39] With this study on the pressure effect on thermal expansivity, they were able to derive geophysically relevant parameters such as the bulk moduli and their temperature derivative. Their experimental results support those calculated from thermodynamic data.

The equation of state was determined for solid hydrogen and deuterium by R. J. Hemley and colleagues who found that the material was more compressible than previously thought.[40] New calculations with single-crystal, X-ray diffraction data to 265 kbar resulted in an equation of state that accurately described both the new high-pressure data and the previous low-pressure compression results.

The reliability of equation of state determinations was also of concern because of the state of stress at megabar pressures. The shear strength of MgO to 2.27 Mbar was therefore measured and significant changes were discovered in terms of the degree and character of the elastic anisotropy.[41] After correcting for a shear strength of 110 kbar at 2.27 Mbar, a single equation of state was obtained that was consistent with a variety of other experimental data.

The inner and outer core of the Earth are less dense than expected for pure iron. Possible core components that may account for the decreased density includes hydrogen, silicon, oxygen and sulfur. The formation of iron hydride was studied by reacting iron and hydrogen and it was found to be stable to pressures of 620 kbar.[42] The investigators concluded from the equation of state that a large hydrogen component (>40 mole%) is compatible with the density of the core determined from seismology.

Ferroelectricity

From first principles and with the aid of supercomputers, a theoretical perspective for ferroelectricity in perovskites has been developed by R. E. Cohen.[43] His goal was to develop an understanding that would lead to the prediction and design of ferroelectric materials for applications to optical, electro-optical, and piezoelectric devices. Piezoelectric response was computed for the first time in a ferroelectric $PbTiO_3$ in the late 1990s.[44] The piezoelectric response in $BaTiO_3$ was induced by rotating the polarization, and it was found to be possible to enhance the response by applying an external electric field oblique to the polarization.[45] This mechanism is apparently responsible for the large electro-mechanical coupling found in a new class of single crystal relaxor ferroelectrics, which will revolutionize the use of ferroelectrics to sonar, hydrophones, and medical ultrasonic imaging.

In the paragraphs above a vast array of techniques have been applied to the study of minerals over a very large range of pressures and temperatures. The materials have been exposed to the entire range of radiation from the near-infrared to gamma rays. Brief mention has been made of luminescence, Brillouin scattering, ultrasonic interferometry, neutron scattering, nuclear magnetic resonance, Fermi surfaces, and various X-ray diffraction schemes. The tools of the physicist have been adapted and enhanced dramatically to solve the mineralogical problems. Major new initiatives, particularly in the use of synchrotron radiation, have been undertaken by the mineral physicists. There is still great potential for future advances in understanding the fundamental properties of minerals and rocks in the entire planet.

16

ISOTOPIC GEOCHEMISTRY

Isotopes have different thermodynamic properties and vary as a function of all the geologically significant factors. In the 1940s, H. C. Urey suggested that the fractionation of the isotopes was sufficiently large to be useful in determining the CO_2 contents of ancient atmospheres and the paleotemperatures of the oceans.[1] From this suggestion a large number of applications were generated and the field of isotopic geochemistry arose. Major applications have been made using the isotopes of C, O, H, N, and S at the Geophysical Laboratory in organic, ecological, petrological, ore-deposit, and extra-terrestrial geochemical problems. Only a small sample of those studies can be given here, yet the success in illuminating significant geological problems will be evident.

Carbon

During the conversion of inorganic carbon to living matter there is an isotope effect.[2] To quantify the effect, P. H. Abelson and T. C. Hoering cultured organisms and extracted various organic portions for isotopic analysis with the mass spectrometer.[3] Most of the portions were depleted in ^{13}C compared to the source carbon. The depletion was explained in part by the composition of the respired CO_2 (Figure 16.1) of *Chlorella pyrenoidosa* released in the dark following exposure to light. Organic carbon thus has a lower ^{13}C to ^{12}C ratio than the carbonate or carbon dioxide of the inorganic environment. The variation in isotopic composition revealed new information on the physiology of a living organism in which the flow of C splits into two or more fractions or streams.

In view of the variations that Abelson and Hoering found within various organic components in the same organism, P. L. Parker wanted to know what the variations might be in a closed system for which all the carbon is derived from the bicarbonate in seawater. He chose a shallow marine estuary near Port Aransas in Texas where there was an abundance of biological samples.[4] Although there was a wide isotopic variation for different species, the individuals of the same species were constant. For this reason, the ^{13}C at the natural abundance level could serve as a tracer in ecological systems. The organic material in the sediment could have been supplied by the biological

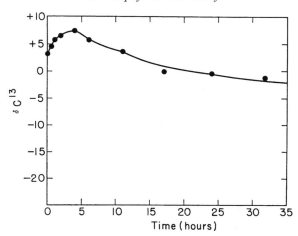

Figure 16.1 The ¹³C content of carbon dioxide respired by *Chlorella pyrenoidosa*. From P. H. Abelson and T. C. Hoering, "The biogeochemistry of the stable isotopes of carbon," *CIW Year Book* 59 (1960), 164, figure 77.

community. In addition, he noted a 5‰ diurnal variation in ¹³C of the inorganic carbon of the seawater. Parker cautioned that because of the variations in isotopic composition of individual species, the sedimentary carbon could not be used as an indication of the history of the sediment.

The isotopic effects in fresh water were examined by M. F. Estep (Fogel) and S. Vigg in two contrasting lakes in northern Nevada, where hatchery-reared fish could be compared with those in the wild.[5] From the water, sediment, and biological samples they showed that the hatchery fish were similar in isotopic content to their artificial food and distinct from their wild species. There was a distinct difference in the isotopes of muscle tissue and scales that could actually be used to trace the fish after release. In summary, the isotopic abundances yielded information in seasonal variations in diet, particularly specific versus generalized feeding and on species specific assimilation and metabolic differences.

Carbon isotopes are one of the few parameters that are preserved in Precambrian rocks, particularly the stromatolites formed by blue-green algae. Their very low ¹³C values (δ¹³C = −10 to −18) indicate that something other than inorganic processes may be needed to explain the differences. A suite of algae and bacteria from a recent hydrothermal environment in Yellowstone National Park were studied by Estep (Fogel) because the extremes of temperature and acidity restricted the number of organisms and the area was undisturbed by man.[6] In addition, the hot springs provided gradients in temperature, pH, dissolved gas content, and species composition. Although the CO_2 levels were close to those in the Precambrian, only a slight increase would mimic the ancient atmosphere. She concluded that biological processes shift the isotopic composition of the photosynthetic bacteria and algae to that equivalent to the Precambrian stromatolites. She found that the most negative carbon isotopic compositions were from blue-green algae that grew in high inorganic C levels, elevated temperatures, and a neutral pH.

Large negative $\delta^{13}C$ values ($\delta^{13}C = -26$ to -13) were also discovered by E. F. Duke and D. Rumble in graphite from plutonic rocks in New Hampshire.[7] Two textural varieties were described wherein the flakes, crystallized at an early stage, represent xenocrystic material assimilated from metasedimentary wall rocks also with large negative values of $\delta^{13}C$. The fine-grained aggregates of graphite with lower negative $\delta^{13}C$ values, on the other hand, precipitated secondarily in a cooling, retrograde hydration stage. Duke and Rumble were not able to identify the source of the fluid or its carbon isotope composition, but they noted that the flakes did not equilibrate with the evolving retrograde fluid. Rumble and T. C. Hoering thought the flakes may have been produced during devolatilization of pelites during metamorphism and the fine-grained graphite precipitated from fluids associated with siliceous carbonates.[8]

In another case of suspected metamorphism due to devolatilization and fluid flow, Rumble and colleagues found no significant shift in $\delta^{13}C$ during metamorphism,[9] but four generations of vein formation resulted in decreasing $\delta^{13}C$. None of the fluids that gave rise to the veins appeared to have achieved exchange equilibrium with the wall rocks.

The isotopic composition of individual carbon molecules can be followed from the diet consumed by an animal into its bones. In an effort to assess the diet of ancient animals it was necessary to determine the effects of diagenesis on the bones. P. E. Hare and colleagues first studied the fractionation between the various organic constituents of modern bones of pigs reared on specific diets.[10] The isotopes among the amino acids, for example, were compared with those from other animals such as lion, lynx, zebra, and whale. The amino acid patterns were consistent even though absolute values differed, even when comparing fossil to modern carnivores. It was established by N. Tuross and colleagues that fossil bone, when well preserved, retained the $\delta^{13}C$ of collagen.[11] Thus, tools are at hand for detecting diets of organisms of ancient times.

Oxygen

As the dominant constituent of rocks and waters, oxygen isotopes have contributed greatly to recording their conditions of formation, alteration, interaction, and equilibration. The oxygen is removed from the rocks or mineral with BrF_5,[12] reacted with carbon to form CO_2, and the isotopes

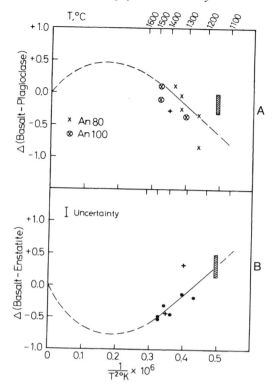

Figure 16.2 Comparison of the ^{18}O content between basaltic melt and plagioclase (A) and basaltic melt and enstatite (B) as a function of temperature. Shaded boxes are phenocryst-groundmass fractionations observed in basalts. Crosses denote exchange with O_2; all others showing the exchange with CO_2. From K. Muehlenbachs and I. Kushiro, "Oxygen isotope exchange and equilibrium of silicates with CO_2 and O_2," *CIW Year Book* 73 (1974), 235, figure 17.

determined with a mass spectrometer equipped with a triple collector detection system. The results are usually expressed in terms of the difference in $^{18}O/^{16}O$ between the sample and Standard Mean Ocean Water (SMOW). Although the pressure effect up to 20 kbar was considered small,[13] the temperature effect (Figure 16.2) was demonstrated by K. Muehlenbachs and I. Kushiro.[14]

The oxygen isotopes of coexisting quartz and magnetite from the Clough formation, Black Mountain, New Hampshire, were measured by Rumble in 1978.[15] He was able to demonstrate that individual layers in the regionally metamorphosed rocks were isothermal within the limits of precision of measurement. Furthermore, the metamorphic fluid varied in its content of H_2O, H_2 and O_2 from layer to layer; that is, the fluid was of heterogeneous composition. Apparently, the mineral assemblages in each layer had sufficient buffer capacity to preserve the premetamorphic heterogeneity of the volatile components.

The fluids were also the focus of a study by C. P. Chamberlain and Rumble in 1988 of the sillimanite zone of metamorphic rocks in New Hampshire.[16] Hot spots were mapped on the basis of the occurrence of granulite facies within the sillimanite zones and networks of quartz-graphite veins. An oxygen isotope alteration halo surrounded the hot spots where metamorphic fluids were focused through a fracture system now recorded by the vein network. The heat transfer by fluids clearly dominated locally over the conductive heat transfer in regional metamorphism.

The values of $\delta^{18}O$ from deep-seated rocks, such as peridotites and xenoliths, are fairly uniform between +5.5 and +5.8‰. In contrast, the acidic igneous rocks, such as granite and rhyolite, have values of +6 to +12‰.[17] It was suspected that the higher values in granite were the result of hydrothermal alteration[18] or assimilation of sedimentary wall rocks.[19] The oxygen isotopes of hydrothermally altered, synmetamorphic granitic rocks from south-central Maine were measured and found to be high, up to 13.8‰.[20] By comparing rocks from different depths, the investigators concluded that the increase was indeed due to the exchange of isotopes between rock and metamorphic hydrothermal fluids.

In contrast, unusually low $\delta^{18}O$ values were found in ultra-high-pressure metamorphic rocks from the Sulu Terrain, eastern China.[21] The values in the eclogite layers ranged from −10.4 to −9.0‰ and those in the quartzites were from −10.2 to −7.3‰. The investigators were of the opinion that these values were acquired through near-surface meteoric water-rock interactions before metamorphism and retained after the rocks had been subducted to great depths and the ultra-high-pressure assemblages formed. Low values of $\delta^{18}O$ between +2.9 and +4.4‰ were also found in ultra-high-pressure, garnet-bearing mafic and ultramafic rocks from Dabie Shan, China.[22]

The coesite inclusions in the omphacite were noteworthy in documenting the peak metamorphism. This was in agreement with the findings from the Sulu Terrain regarding when and where these values were imposed on the rocks. The study involved the use of the new laser microanalytical technique devised by Z. D. Sharp.[23] The advantage of analyzing a small area has greatly expanded the opportunities to unravel the thermal and hydrological history of rocks.

The use of oxygen isotopes has also been of great benefit in resolving organic problems. All the free O_2 produced by photosynthetic organisms is in turn consumed in large portion by other organisms; that is, the Earth's atmosphere reflects global compensation. About 60 percent of the global O_2 consumption is due to various plant processes and the remaining 40 percent is due to nonphotosynthetic organisms such as microbes. Relative to the $^{18}O/^{16}O$ of seawater (SMOW) the atmosphere is $+23.5‰$, which is termed the Dole effect.[24] Guy and colleagues were concerned with the oxygen fractionation during plant respiration in which the only source of oxygen was water with $\delta^{18}O = 0‰$. They found that within several hours, asparagus mesophyll cells were emitting $\delta^{18}O_2 = 21.5‰$, close to that in normal atmosphere.[25] In an expanded study they tried to account for the subtle difference, but concluded that enzymes, dark respiration, or other discriminating pathways may account for the remaining 8.5 percent difference between the Dole effect and the $21.5‰$ of asparagus mesophyll cells.[26]

Ore geologists have long been intrigued by the occurrence of sedimentary manganese depositions. Conditions that may have influenced the deposition are the oxygen content of the water and microbial catalysis. For this reason the $\delta^{18}O$ was measured for manganese oxides and manganates prepared with different water sources having distinct $\delta^{18}O–H_2O$ values while keeping the $\delta^{18}O–O_2$ constant. In addition, the manganates were prepared in three different ways: (1) abiotically, (2) biologically with oxidizing dormant spores, and (3) biologically with active cells of oxidizing marine bacterium, all at pH = 7.6 and temperatures of 20–25 °C. Significant incorporation of molecular O_2 by a single step process of Mn^{2+} to Mn^{4+} was found.[27] About 32–50 percent of the oxygen in the minerals prepared chemically and biologically was derived from molecular oxygen in accordance with many natural environments. Those prepared with dormant spores exhibited little fractionation, and were similar to those in a manganese nodule from Oneida Lake, New York. On the other hand, the manganates from the Kaikata Seamont appear to reflect the $\delta^{18}O$ of seawater exclusively. The conclusion was reached that as a result of different pathways, the manganates may not always reflect a dissolved oxygen signal.

Hydrogen

Plants grown under controlled conditions fractionate hydrogen isotopes reproducibly. Whereas photosynthesis is the major process, additional fractionation occurs in the dark.[28] Little hydrogen isotope fractionation appears to take place during aerobic respiration. The data provide a basis for interpreting isotopes in ancient organic matter.

The hydrogen isotopic ratio was found to be valuable in tracing the food web not only in plants but also in animals. Estep (Fogel) and Dabrowski measured the isotopes in a natural population of snails from Boothbay Harbor, Maine, and in laboratory-reared mice.[29] It was the hydrogen isotopic content of the food, not the water, that determined the result. The snails had fed exclusively on brown algae and inherited their D/H, not that of other available algae. Similarly, the liver and muscle tissue of the mice had the same ratio as their food and not the water. They came to the same conclusion based on ^{13}C measurements. Further support of the concept was obtained by Macko and colleagues.[30]

More detailed experiments on hydrogen isotope fractionation were undertaken by Estep (Fogel) and Hoering on the growth of microalgae involving light and dark reactions of photosynthesis.[31] Growth in red and white light activated phosphoglyceric acid reduction and hydrogen isotope discrimination when H was fixed into organic matter. Additional fractionation of -30 to $-60‰$ occurred during biosynthesis of proteins and lipids and was associated with glycolysis. The complex network of reactions between H_2O and the organically bonded hydrogen was, in their view, regulated principally by photosynthesis, whereas secondary regulation took place during glycolysis. It was clear that stable isotope measurements could be used to identify the key reactions in plant metabolism.

Sulfur

The lead isotopes of galena, PbS, were determined by Doe in 1962[32] to obtain the radiometric age of granites and pegmatites, and a dozen investigators measured the sulfur isotopes of coexisting natural sulfides of complex composition. To ascertain the fundamentals, Puchelt and Kullerud focused on the pure Pb-S system for which the phase relations had been determined and the compositions of the phases were simple.[33] The sulfur isotope ratio, $^{34}S/^{32}S$, in galena exchanged rapidly with a sulfur vapor phase even at 300 °C. Consequently, the ratio would not be retained over geological time if exposed to sulfur-bearing gases. Pronounced post-depositional zoning is to be expected. Because of the difference in reaction rates, the sulfide is enriched in the light S isotope. When equilibrium was obtained in the experiments at 300 °C, the ratio was 2.5‰ and trailed by three orders of magnitude the achievement of phase equilibrium. For this reason, it is not likely that the original isotope ratio will be ascertained in natural systems.

The stable isotopes of sulfur were relatively uniform in ancient sedimentary rocks before 2.8 billion years ago and exhibited wide dispersion after that date. The change was attributed to the onset of sulfur-reducing and oxidizing bacteria. To test this conclusion, Hoering collected some 3 billion-year-old barite in India[34] and used a new method[35] to measure all the stable isotopes of sulfur. The value obtained for $\delta^{35}S$ (4.26‰) agreed with other samples of the same age, but the contained detrital pyrite raised a critical question. How did authigenic sulfates of Archean age arise? Non-biological oxidation by atmospheric oxygen may have been responsible, but pyrite would have been readily oxidized. The occurrence of this ancient barite, which has not been enriched in the heavy isotope, remains an open question.

The biological issue arose again in samples from a borehole in the Creede Caldera, Colorado. A wide range of sulfur isotope variations from $-18‰$ to $+70‰$ were found using a new technique, the infrared laser microprobe.[36] Spatially resolved analyses record isotopic zoning that becomes progressively enriched in ^{34}S with decreasing depositional age in individual veins and stalactites. The interpretations led to two hypotheses: first, that sulfur originated as volcanic H_2S and the pyrite precipitated by reactions with iron-bearing minerals; and second, that reduction of sulfate to sulfide by anaerobic bacteria was catalyzed by enzymes. Although the researchers did not think they had a complete database for choosing between the two hypotheses, they favored the latter hypothesis.

Some isotopically light pyrite and pyrrhotite, $\delta^{34}S = -27‰$, were found in some black shales in Maine.[37] Apparently the metamorphism grossly disturbed the rocks isotopically. Lack of equilibrium between the rocks and an infiltrating fluid was believed to occur because of channeling of fluid flow along rather than across layers. The predominance of the isotopically light isotopes in graphitic schists is consistent with substantial preservation of premetamorphic values reflecting anoxic black shale depositional environment. Nevertheless, the investigators caution that preservation of isotopic compositions should not be routinely taken as evidence of "fluid-absent" conditions or internal buffering of fluid composition.

Nitrogen

The most abundant element in the Earth's atmosphere is nitrogen, which occurs in two stable isotopes, ^{14}N and ^{15}N. It has also been found in volcanic gases as free N_2 and as NH_3, in both fresh and salt water as NH_4^+, NO_3^- ions, in meteorites, and especially in soils. It is alleged to be a constituent of all living matter, but it has not been established as a requisite element at the early stages of the origin of life. Some organisms can assimilate N from the air, others obtain it from inorganic compounds in soil, some micro-organisms convert ammonium salts to nitrates, and in turn some bacteria decompose nitrogen compounds and return the element to the air. Many questions are unanswered in regard to the initial source of N on Earth and its retention during the early stages of the Earth's history. The current need is for extensive studies on the role of nitrogen in biology and the Geophysical Laboratory is attempting to meet some of that need.

The variation of N in naturally occurring substances was measured by Hoering in the mid-1950s, who used the N_2 in air as the standard.[38] He found values of $^{15}N/^{14}N$ ranging from $+13‰$ in natural ammonium chloride to $-13‰$ in methane from a gas well. Marine sediments, animals, and plants all have $\delta^{15}N$ greater than $0‰$.[39] In most areas of the ocean, nitrogen is a limiting nutrient and determines the amount of growth. Nitrogen occurs in many forms and oxidation states in seawater, but NH_4^+ appears to be the preferential form utilized by marine phytoplankton.[40]

Sediments in near-shore coastal areas derive their organic matter from both terrestrial and marine sources. Identification of the origin of specific organic materials has been possible by N isotope-ratio analysis.[41] Three coastal areas of the USA were studied: Chesapeake Bay, Delaware Bay, and New York Bay. On the basis of $\delta^{15}N = 0\%o$ as indicative of a terrestrial source and $+ 6.2\%o$ as indicative of a marine derivative, the relative proportion of the sources were identified. The Delaware Bay analysis indicated the major organic input was of planktonic origin. The New York Bay sediments with inputs from the Hudson River were diluted with a marine component as the distance from shore increased. The Chesapeake Bay sediments, although exhibiting the general relationship of the other systems, were enriched in the marine source in the central part of the bay and had higher contents of organic matter. Many processes can influence such results and several have been examined in detail.

Microbial alteration of biochemical substances is a major process in the early diagenesis of organic deposits.[42] Diverse populations of bacteria serve as food sources. Cultures utilizing particular amino acids may be depleted in ^{15}N. Large negative isotope fractionation may result from loss of ^{15}N-enriched ammonia or fractionation of ammonia during glutamate synthesis. Bacteria grown on aspartic and glutonic acids were enriched in ^{15}N relative to the substrate. With such diversity, large isotope fractionations could cancel each other and not be observed. On the other hand, some organic material could be altered extensively before it is polymerized into microbiolgically resistant humic acids and kerogen. These conclusions have since been substantiated by experiments on cultured bacterial.[43]

The changes that take place in organic materials after burial are also reflected in the fossil record. The collagen from both modern and fossil bone have been studied by P. E. Hare and M. Estep (Fogel).[44] Typical amino acid patterns for collagen were exhibited by both samples. The non-essential amino acids synthesized by the animal itself in modern bone were close to that of the total protein, whereas the essential amino acids required in its diet showed a wide range of values. The same isotopic signatures were still present in the fossil bone some 10,000 years old. The investigators were confident that the stability and significance of organic matter in fossils can be determined by a combination of amino acid geochemical and stable isotope techniques.

One of the most important pursuits of the organic geochemist is how organic material breaks down at low temperatures into more stable products over geological time. Hoering has examined how kerogen, which comprises the bulk of organic matter in sedimentary rocks, decomposed successively.[45] Because the intermediates formed are constituents of petroleum, it was indeed a critical study. Previous studies on the pyrolysis of kerogen in a dry inert atmosphere yielded large amounts of olefinic hydrocarbons rare in petroleums, but when carried out in excess liquid water, no olefins were found and the products were close to those in petroleum. To check the role of water, Hoering added heavy water, D_2O, in order to learn if water actively participates in the reactions. Using a natural shale, he held the sample in a pressure vessel at 330 °C for three days. Extensive substitution of deuterium occurred. The reaction products from the shale included many of the same compounds that occur in old petroleums. Hoering noted that his studies involved only one of the variables that control the breakdown of organic materials, but believed continued use of isotopic and molecular probes would resolve the question.

High concentrations of NH_4^+ in igneous and metamorphic minerals indicate that N may be an important fluid constituent. Nitrogen is a common constituent in volcanic gases, in hydrothermal deposits, fluid inclusions, and in the sediments described above. The NH_4^+ readily substitutes for K in the feldspars and micas. Although N may be strongly depleted in metamorphism, there is a shift toward higher $\delta^{15}N$ with increasing metamorphic grade. G. E. Bebout and M. L. Fogel examined in detail the N data for the Catalina schists of California with progressive devolatilization and its impact on the mobility of other trace elements with progressive volatile loss.[46] There appears to be efficient retention of N during diagenesis and low-grade metamorphism. There is a progressive partitioning of N into H_2O-rich fluids generated by the breakdown of mainly chlorite and white mica with increasing grade of metamorphism. The shifts in $\delta^{15}N$ and N concentration in the Catalina schists are consistent with N_2-NH_4 exchange as the dominant mechanism of N-isotope fractionation. Bebout and Fogel believe that the N-isotope system is also a valuable measure of open versus closed system behavior during metamorphism. The isotopes are clearly a measure of large-scale volatile transport.

Hydrothermal environments provide another setting that affect the cycling and isotopic composition of nitrogen through organic matter. Estep (Fogel) and S. A. Macko reported on the stable isotopes in hydrothermal areas of Yellowstone National Park to uncover the factors that govern $\delta^{15}N$ in evolutionarily primitive organisms.[47] In the open system, hot springs there are higher nonlimiting concentrations of combined N, and the organic matter reflects the normal isotope fractionation associated with enzymatic reactions of biological metabolism. In closed system pools, however, the isotopic composition is distinctly different and indicates the importance of internal recycling of ammonia. Aside from the nitrogen-fixing organisms such as blue-green algae, nonphotosynthetic bacteria occur at higher temperatures up to 60 °C. In addition, bacteria that derive their cellular energy from nitrification or denitrification contribute to the variability. The presence of distinctive pathways for N uptake can thereby be identified and the presence or absence of particular groups of micro-organisms assessed even in thermophilic environments.

From the above brief presentations, one can appreciate the great value of determining the isotopes of C, O, H, S, and N from a wide range of environments. It should also be evident that there is a great need for continuation of the investigation of simple, basic systems as well as defining the complex natural systems.

17

GEOCHRONOLOGY

Geology is a historical science highly dependent on a time scale. Initially the fossil record was used to document the sequence through changes of primary morphological features. Estimates of sedimentation ratio, the Earth's cooling, and salt accumulation in the ocean, were also used to construct a timetable in the beginning. After the discovery of radioactivity,[1] the genetic connection between radium and uranium was derived from an examination of ancient minerals.[2] Ernest Rutherford, in 1905, was the first to compute the age of a mineral from measured U-He abundances.[3] The age of a mineral based on the abundances of U and Pb were first calculated by B. Boltwood two years later.[4] In 1928, C. N. Fenner[5] used new analytical methods for the determination of uranium, thorium, and lead as a basis for age calculations. In the following year, C. N. Fenner and C. S. Piggot enlisted the help of F. W. Aston at Cambridge, England[6] in making the first calculation of the age of a mineral (thorian uraninite) on the basis of the specific isotopes of lead determined by mass spectroscopy. The discrepancy between the Pb-U age and Pb-Th age, however, led them to believe that the assumed U-Th equivalence factor may be in error. Although the radioactivity of ocean sediments became a major interest of Piggot and W. D. Urry,[7] the age determination of minerals was dominated by others until after World War II.

A program for the determination of the age of minerals was initiated as a collaborative effort with the Department of Terrestrial Magnetism (DTM) in 1950. The goal was to develop techniques and equipment for determining the age of several common minerals in the same rock for which different methods could yield completely independent ages. Granites of Precambrian age were chosen from which almost a dozen major and accessory minerals were extracted. The methods focused on the accurate measure of the naturally occurring radioactive elements having long half-lives. The parent and daughter elements were concentrated from each separated mineral by ion-exchange resins after spiking the sample with a known amount of a tracer isotope and digesting it in acid. In this way the absolute concentration of the isotopes in the mineral could be ascertained by analysis in a mass spectrometer. The mineral separation and solution chemistry was performed by G. L. Davis at the Geophysical Laboratory and the mass spectrometry was carried out mainly by others at DTM (L. T. Aldrich, G. W. Wetherill, and G. R. Tilton).

In 1956 Tilton transferred from DTM to the Geophysical Laboratory, and the close cooperation of both groups, with a growing tide of postdoctoral fellows and guest investigators, was even more effective.

Between 1950 and 1955 efforts were concentrated on methods. With the increased sensitivity of the isotope dilution technique, the decay scheme of ^{87}Rb to ^{87}Sr was applied successfully to the Li-bearing micas. The ^{40}K/^{40}Ca and ^{40}K/^{40}Ar clocks were slowly being developed.[8] The discrepancies between the various clocks were attributed to radiation damage in the zircons, differential leaching in acids, and the transfer of Pb between minerals.[9] These problems were investigated in great detail. The highest ages appeared to be given by the Rb-Sr clock, whereas the lowest age was found by the Th-Pb method. It was deduced that the Rb half-life used in the calculation was apparently too low, so a new half-life was found by assuming the U-Pb age was correct in a mineral giving concordant ages from six locations having a range of ages. Values of the half-life were calculated from the ^{87}Rb/^{87}Sr from Rb-bearing minerals at these same localities. The new Rb half-life so determined geologically, eliminated some of the previous discordance and it remains the present-day value.

With improved techniques, it became possible to begin applying the results to the solution of geological problems. In 1957, after the discovery of large groups of ancient rocks, Tilton and colleagues developed the concept that the oldest rocks were the "nucleus" of a continent and younger belts of rocks were subsequently added on.[10] Because of the different responses of specific minerals to metamorphism, it became possible to identify the age of the critical elements in the geologic history of a region. In 1958 a major problem among discordant age values was resolved. Two years earlier Wetherill had shown in a plot of ^{206}Pb/^{238}U vs. ^{207}Pb/^{235}U that the various measurements could be related by a line of "concordia."[11] The line was interpreted by Tilton in 1958 as the locus of apparent ages resulting from the continuous loss by diffusion of Pb (Figure 17.1).[12] It was this discovery that cleared the way for geological application on a grand scale. Subsequently, the zones of various ages mapped for the USA and Canada were confirmed, and the concept of the slow accretion of a continent began to take shape.

Discordant ages continued to plague the analysts, but each time the cause of the discordancy was resolved greater insight into geologic processes emerged. Studies of contact aureoles and eventually regional metamorphic grades illustrated that temperature affected the diffusion of radioactive daughter elements. The apparent ages could then be used for mapping thermal zones in a metamorphosed region.

When the amphiboles and pyroxenes became useful indicators of age, a new range of petrological problems could be tackled. For example, the exceptionally low concentration of U in dunites and websterites suggested that the heat flow in the oceans had to arise from another source rock;

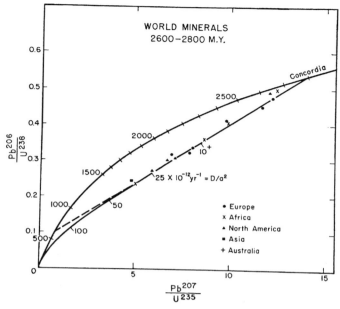

Figure 17.1 Concordia: Curve formed by plot of $^{206}Pb/^{238}U$ vs. $^{207}Pb/^{239}U$ whose values increase in value due to the nuclear decay of uranium to lead with passage of time, assuming a closed U-Pb system. If the system is not closed, the curve through the points showing discordant ages is calculated for the loss of lead by continuous diffusion from spheres of radius *a* having an age of 2800 million years. The data are interpreted as evidence of either a loss of lead 600 million years ago, from minerals crystallized 2800 million years ago, or alternatively, as continuous diffusion of lead from the crystals. From G. R. Tilton, "Volume diffusion as a mechanism for discordant lead ages," *J. Geophys. Res.* 65 (1960), 2393, figure 1. By permission of the American Geophysical Union.

eclogite appeared to be adequate. On the other hand, the hornblende-bearing peridotites in some of the islands had enough K to yield the appropriate heat flow. In addition, the isotopic analysis of Pb indicated that the basalts of the oceanic islands were from heterogeneous sources.

With the arrival of T. Krogh in 1966 the major thrust of the work turned to the Grenville controversy. The value of analyzing the whole rock as a closed system instead of individual minerals became appreciated. As belts of ages were identified, the region appeared to have analogues with the modern-day, island-arc volcanic zones. The Grenville front was interpreted as an ancient plate boundary where a major metamorphic event took place 1500–1800 million years ago with a second major dislocation about 1000 million years ago.[13]

In the next ten years (1968–1978) at least three major improvements in technique took place. In 1973, T. E. Krogh invented a new dissolution method for zircons in which a teflon-lined pressure vessel was used at 220 °C.[14] X-ray fluorescence became a standard tool for ascertaining the suitability of samples for Rb-Sr analysis. The production and purification of the ^{205}Pb spike (with the help of the Holifield National Laboratory)[15] greatly improved the precision of the clocks based on lead. The use of the ^{205}Pb spike was apparently developed independently at the same time by F. Tera and G. J. Wasserburg.[16] With these improvements, the ages of zircons in kimberlites could be measured.[17] The African diamond pipes were found to be around 90 million years old and two groups of pipes at Yakutsk, USSR, were 402–443 million years and 360–344 million years. Other studies included the dating of many other geologically significant formations.

The use of the racemization of amino acids as a dating method[18] has been described in Chapter 11.

With the resignation of T. E. Krogh in 1975 and the retirement of G. L. Davis in 1978, the geochronology program at the Geophysical Laboratory was re-evaluated. It was evident that the Laboratory had served its role in pioneering new methods and contributing new concepts to the solution of geologic problems dependent on knowledge of accurate isotopic compositions. The existence of more than fifty laboratories in the USA devoted to the dating of rocks and minerals, many headed up by past associates, indicated that the field was well established, and it was appropriate for the Laboratory to invest its limited resources in new opportunities.

18

ELEMENT PARTITIONING

The relationships among rocks are primarily deduced from the assemblage of minerals. Because of solid solutions, alterations, and nonequilibrium effects, only general relations are usually attained. With accurate knowledge of the composition of the phases in the assemblage, however, the specific conditions of temperature and pressure of formation can be determined as well as the state of equilibrium and extent of alteration. At present, the specific conditions are estimated from the element partitioning between at least two phases. The partitioning of Ca-Mg, Fe-Mg, Fe-Ti, Fe^{2+}-Fe^{3+}, Ti-Al, and Ni-Cu, for example, have been calibrated in terms of pressure and temperature. The partitioning between a crystal and liquid, volatiles and liquid, as well as between immiscible liquids, have been investigated. The rare earths and other trace elements were measured successfully where appropriate analytical techniques were available. With calibration, the partitioning of elements has been very useful in modeling the formation and evolution of rocks.

Ca–Mg partitioning in pyroxenes

One of the most useful geothermometers arose from the partition of Ca and Mg between diopside and enstatite. Because of rate problems the earliest studies on the pyroxene join showed little or no change in partitioning with temperature.[1] With the use of a flux, LiF, a large change in partitioning was observed,[2] and a solvus demonstrated in accord with that seen in natural pyroxenes.[3] The solvus was refined by F. R. Boyd and J. F. Schairer in 1964[4] at one atmosphere (Figure 18.1) and determined at 30 kbar by B. Davis and Boyd in 1966.[5] Although a small pressure effect was seen, the latter authors concluded that "these phase relations provide a potentially useful geothermometer, which is relatively independent of pressure" (p. 3574). Ten years later, D. H. Lindsley and S. A. Dixon investigated the solvus between 5 and 35 kbar, and declared that there is "clearly a pressure effect," and it increases with increasing temperature.[6] In addition, they said the solvus was not affected by the presence of excess SiO_2 or $MgSiO_3$. Although the pressure effect may lead to small errors, the influence of other components such as Fe^{2+} were of greater concern. It was Boyd's opinion that the solvus was useful provided the diopside contained less than 10-mole% $FeSiO_3$, the limit

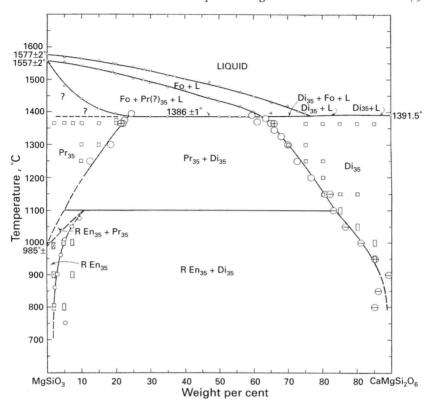

Figure 18.1 Equilibrium relations on the join $MgSiO_3$-$CaMgSi_2O_6$ illustrating the two-pyroxene solvus and the partitioning of Ca-Mg between phases. From F. R. Boyd and J. F. Shairer, "The system $MgSiO_3$-$CaMgSi_2O_6$," *J. Petrol.* 5 (1964), 280, figure 1. By permission of Oxford University Press.

in most lherzolite nodules.[7] The pyroxene solvus continues to serve as a major geothermometer.

Mg–Fe partitioning

The ratios of $Mg/(Mg + \Sigma Fe)$ of magmatic liquids are often used as a measure of their origin by the partial melting of a peridotite. B. O. Mysen used this ratio in olivine to ascertain if it is in equilibrium with its coexisting liquid provided the pressure, temperature, and fO_2 could be determined.[8] On the basis of his analysis, he concluded that the present data were insufficient to justify correlations of the Mg/Fe content of olivines with their containing liquid. Another factor was revealed by Mysen and I. Kushiro when they found that partial melts were compositionally controlled by their coexisting silicate minerals.[9]

A more specific approach to the Mg-Fe partitioning in clinopyroxene was undertaken by McCallister and colleagues.[10] They determined the distribution of Mg and Fe^{2+} between the M1 and M2 sites as a function of temperature. The crystallization and subsolidus cooling history were deduced for some natural pyroxenes after complete crystal structure refinements were carried out on each crystal after experiments at various temperatures.

The partitioning of Mg-Fe in coexisting clinopyroxene and garnet was used to classify natural eclogites[11] and it was inferred that the differences were due to different P–T conditions. The concept was applied as a geothermometer by Mysen and K. S. Heier in 1972 for the eclogites of Hareidland, Norway.[12] The partitioning was calibrated experimentally by A. Råheim and D. H. Green in the range 600–1500 °C and 20–40 kbar.[13] It was the opinion of A. Finnerty and Boyd that the Mg-Fe exchange reactions were less precise because of the lack of corrections for Fe^{3+}, but they were confident that the thermal gradients could eventually be ascertained with that method.[14] The influence of Fe^{3+} on the Mg-Fe geothermometer was investigated by Luth and colleagues.[15] They found, for example, that the presence of ∼12% of Fe as Fe^{3+} in garnets increased the calculated temperatures by >200 °C. In addition, there was a concomitant increase in pressures of ∼10–15 kbar calculated from geobarometers involving the Al content in orthopyroxene (see below). They demonstrated that Fe^{3+} could not be neglected in consideration of the application of geothermometers. Furthermore, they believed it was incorrect to correlate the Fe^{2+}/Fe^{3+} in garnets directly with the fO_2 of the source region without taking into account other compositional parameters.

Al in enstatite, a geobarometer

Equilibrium pressure may be estimated from the Al_2O_3 content of pyroxene crystallized with garnet. If the pressure is increased at constant temperature on a pyroxene-garnet assemblage, the Al_2O_3 content of pyroxene decreases and more garnet is formed. This effect was first predicted by A. E. Ringwood in a personal conversation to Boyd (recorded in 1973[16] and first demonstrated in the synthetic system by Boyd and J. L. England[17] and by I. D. MacGregor and Ringwood with analyzed natural samples[18]). The Boyd and England calibration was determined at two isotherms (1100 °C and 1600 °C) at 30 kbar and extended by MacGregor from 900 °C to 1600 °C at pressures from 5 to 40 kbar.[19] When applied to natural assemblage, errors were anticipated from other compositional variables. The effect of TiO_2 and FeO, for example, affects the solubility of Al_2O_3 in orthopyroxene,[20] and results in estimates of equilibrium pressure that are too high. Nevertheless, the aluminum content of enstatite coexisting with garnet and clinopyroxenes is the most widely used thermobarometer for garnet lherzolites.[21]

Ni–liquid partitioning

The great economic importance of nickel has focused considerable attention on the partitioning of that element in magmas.[22] Because Ni is essentially a trace element, the analytical problems of determining its content in silicate phases is substantial. The problem was resolved to a large extent by Mysen in the mid-1970s, who used ^{63}Ni and determined its content by mapping β-tracks in nuclear emulsions.[23] He measured initially the partitioning of Ni in olivine and aluminum enstatite coexisting in a hydrous liquid of haplobasaltic composition at 1025 °C at 20 kbar. With additional experimentation Mysen concluded that the partitioning was dependent on the nickel concentration as well as temperature, pressure, and proportions of the coexisting phases.[24] In short, the use of nickel partitioning to determine if a rock is a direct melting product or the result of a fractionation product will require measurement of a wide array of variables. By simplifying the system to just olivine and an anhydrous melt, Mysen found the partition coefficients were independent of the Ni content until 1000 ppm in Ni was exceeded.[25] Fractional crystallization of the olivine results in a lowering of the Ni content of the magma, but the amount depended on the nickel content of the magma and the pressure, which probably induces structural changes in the silicate melt. The structural changes were specifically noted by Mysen and Kushiro in regard to the changing role of Al in silicate melts.[26]

The influence of melt structure on trace-element partitioning was emphasized in a study by Mysen and D. Virgo.[27] They concluded that a few weight percent of a minor oxide can have discernible effects on the partitioning, but may not be petrologically significant. On the other hand, when all oxide concentrations are considered, the partitioning may be significantly affected by liquid composition.

REE partitioning

The first of the fourteen rare earths to be studied experimentally for its distribution factor was cesium. In the 1950s H. P. Eugster grew sanidine and phlogopite in a hydrothermal environment containing known amounts of the rare element (Figure 18.2).[28] Although only one pressure (1 kbar) was investigated, it was clear that the rare earths would be a sensitive measure of the conditions of crystallization. His next step was to compare the Cs/K in sanidine with the coexisting vapor in the range 400 °C to 800 °C.[29] The temperature dependence of the ratio was greater than expected and the surprise was that at 700 °C the Cs content of the feldspar was equal to that

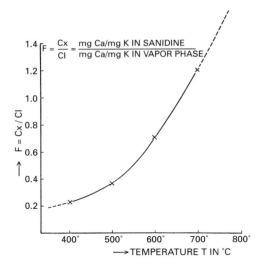

Figure 18.2 Distribution of cesium between sanidine and vapor as a function of temperature at 1 kbar. *Source*: From H. P. Eugster, "Distribution coefficients of trace elements," *CIW Year Book* 53 (1954), 104, figure 3.

of the vapor. Although only two pressures (1 and 2 kb) were studied, Eugster believed the pressure influence was within experimental error.

The β-tracking technique was used by M. G. Seitz in measuring the partitioning of samarium between diopside and a basalt melt at 20 kbar.[30] Because of slow diffusion rates, Seitz was concerned about achieving total equilibrium, but noted that a close approach to equilibrium is obtained when just the surface of the crystal and adjacent glass are measured. He believed the lower partitioning coefficients he obtained compared with those of Maseda and Kushiro[31] reflect his closer approach to equilibrium.

The partitioning of samarium was also investigated by Mysen between olivine and a hydrous silicate liquid,[32] and by B. J. Wood between garnet and a hydrous silicate liquid.[33] They both noted that the partitioning is not only a function of the total REE content, but that pressure, temperature, and bulk composition influence the result.

In order to evaluate the influence of the degree of melting on partitioning, W. S. Harrison found the coefficients for Ce, Sm, and Tm between garnet, clinopyroxene, orthopyroxene, olivine, and coexisting melt of a natural peridotite nodule from 2.3 to 37.7 percent melting.[34] She found that the assumption of constant partitioning was an oversimplification; whereas some changed by a factor of two, others did not change significantly. Although discouraged by the observation that the fractionation was not typical of alkali basalt, which probably requires a metasomatic enrichment,[35] she was encouraged by the fact that simple systems could model the melting behavior of natural systems.

Vapor influence on partitioning

The effect of vapors on REE partitioning between minerals was assessed by Mysen in 1981.[36] He obtained the partition coefficients of samarium in diopside with various proportions of CO_2 and H_2O from 900 to 1100 °C at 20 and 30 kbar. The partitioning into the vapor was so high that he concluded CO_2-rich fluids would act as metasomatizing agents in the upper mantle.

The presence of fluorine and chlorine as essential constituents of apatite also calls attention to those volatiles in affecting partitioning in the presence of both H_2O and CO_2. From a wide range of compositions run at 950 °C and 1050 °C under 20 kbar, J. Brenan concluded that chlorine, in particular, will have an important role in trace element mobilization.[37] Again he warns about drawing conclusions from the composition of apatite about the halogen in the mantle without evaluation of the composition of the entire system.

In the presence of CO_2, R. F. Wendlandt and W. J. Harrison investigated the partitioning of Ce, Sm and Tm between coexisting carbonate and silicate melts at 1200 °C and 1300 °C at 5 and 20 kbar.[38] The two-liquid region covers over 60 percent of sanidine–potassium carbonate join. The rare earths favored the carbonate melt by a factor of two to three for the light REE and five to eight for the heavy REE. They believe their data explain the enhanced REE enrichment of carbonatites, alkali-rich silicate melts, and kimberlites.

The studies described briefly above are only a small sample of the many investigations undertaken in element partitioning at the Geophysical Laboratory. Others include Ti-Al in coexisting silicates and liquids; Fe-Ti in oxides, and P-Si in silicate melts. Another array of valuable partitioning data is to be found in the isotopes of H, C, O, N, and S. Those investigations were presented in Chapter 16 on isotope geochemistry.

19

PETROFABRICS AND STATISTICAL PETROLOGY

The experimental investigation of the deformation of rocks "under conditions of great stress" was one of the primary areas listed by the Advisory Committee in Geophysics for study by the proposed Geophysical Laboratory.[1] While at the US Geological Survey, G. F. Becker had initiated studies in 1893 on "Finite homogeneous strain, flow and rupture of rocks,"[2] and was given a CIW grant to continue the work on the elasticity and plasticity of solids (Grant No. 172). In addition, F. D. Adams at McGill University was given grants (Nos. 4 and 117) for his studies on the "flow of rocks." In 1906, F. E. Wright was able to reproduce the textures of certain metamorphic rocks by crystallizing glass under unequal stress.[3] Unfortunately, this field of study was largely ignored by US geologists until a few recognized the need and traveled abroad for training.

Observations of rock deformation in the field consisted mainly of mapping folds, fractures, faults, platy cleavage, and plastic creep. An example is the classic paper in 1936 by T. F. W. Barth on the structural and petrologic studies in Dutchess County, New York.[4] The laboratory study of the fabric of rocks, developed by Austrian geologists B. Sander and W. Schmidt,[5] was undertaken by E. Ingerson (trained by Sander in Innsbruck), using the powerful new tool of measurement of the orientation of individual grains with a universal stage microscope (Figure 19.1).[6] Hundreds of grains had to be measured and their statistical study was made on an equal area projection net. Ingerson used the new method on a complex biotite-muscovite schist from Niederthal, Tyrol. He advanced "fabric analysis" greatly by elaborating and illustrating each technique correlating the relative ages, intensities, and directions of motion during the two deforming movements involved. Later, Ingerson contributed a major portion of a GSA Memoir on "Structural Petrology" on the laboratory techniques of petrofabric analysis.[7]

The Director, L. H. Adams, replaced Ingerson in 1947 when he left for the US Geological Survey with Felix Chayes who was chosen "for the purpose of conducting studies on the application of certain statistical methods to petrometry." Chayes had been trained at Columbia University under Professor S. J. Shand, who had developed a recording micrometer with a traversing mechanism for measuring relative areas in thin sections.[8] Chayes developed a point counter that provided rapid and precise analysis of thin

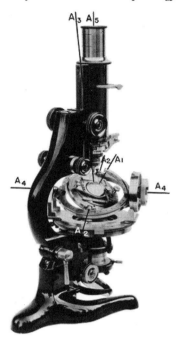

Figure 19.1 Microscope equipped with a universal stage showing the axes off zero to illustrate the graduated circles. From E. B. Knopf and E. Ingerson, "Structural petrology, Part II. Laboratory technique of petrofabric analysis," *Geol. Soc. Am. Memoir* 6 (1938), plate 23. By permission of the Geological Society of America.

sections.[9] He was particularly concerned about the random-sample problem, the definition of a single rock type, and the bias due to grain size. After an extensive collection of twelve New England granites he found the variation in mineral composition to be remarkably small. The largest observed standard deviation for quartz in these samples was 2.9 percent, whereas the range in mean values was from 23 to 33 percent. It was his opinion that the granites involved must have been formed from a homogeneous parent material. This view was strongly supported by the fact that the quartz-potash feldspar-plagioclase values always exhibited only one of the four possible relations to the thermal valley in the system soda feldspar-potash feldspar-silica (Figure 19.2). Even though the pure system is limited in chemical composition, he found no exceptions to the individual plots of the natural rocks in regard to the three-phase boundaries of the synthetic system. The techniques of modal analysis were presented by Chayes in a book that grew out of his lectures at the California Institute of Technology during the winter of 1955.[10]

In the course of his point-counting studies on rocks with potash feldspar, Chayes revised the technique of staining the feldspar for easy recognition in

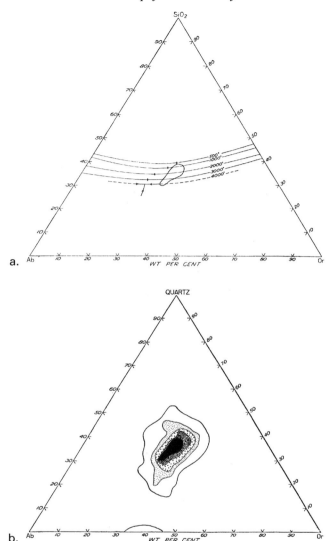

Figure 19.2 Effect of water pressure on the isobaric minimum in the system NaAlSi$_3$O$_8$(Ab)–KAlSi$_3$O$_8$ (Or)–SiO$_2$–H$_2$O. The normative compositions of 571 analyzed plutonic rocks that carry > 80 percent normative Ab + Or + Q, shown solid in 19.2b, is outlined in 19.2a. From O. F. Tuttle and N. L. Bowen, "Origin of granite in the light of experimental studies in the system NaAlSi$_3$O$_8$–KAlSiO$_3$O$_8$–SiO$_2$–H$_2$O," *Geol. Soc. Am. Memoir* 74 (1958), figures 38 and 42. By permission of the Geological Society of America.

thin section.[11] After etching the slide with HF, a dilute solution of sodium cobaltinitrite in water at room temperature was applied for about five minutes and rinsed. The bright lemon yellow color was an immense aid in identifying the potash feldspar and is now widely used.

Some 1800 complete chemical analyses of Cenozoic rocks called andesite were examined statistically by Chayes, and their average values and sample frequency distribution presented in 1969.[12] Because of the explosive nature of andesite eruptions, he particularly noted the average H_2O content of 1.24 percent, but said the observed H_2O content was not a sound basis for estimating the H_2O content of andesite magma since much H_2O may have escaped during eruption or even been added hydrothermally or by surficial weathering.

Chayes continued to collect published analyses of igneous rocks well beyond H. S. Washington's substantial collection published in 1917.[13] He estimated that there are at least 61,834 published analyses of rocks, but realized this was a considerable underestimate. He felt that there had been no serious attempt to identify the sources of error in the classical silicate analysis of rocks. The cooperative investigation of specific specimens of granite (G-l) and diabase (W-l), reported by H. W. Fairbairn in 1951 was a disappointment even though immensely informative.[14] The original G-l sample, which has been used as an international, interlaboratory, geochemical standard since 1949, was collected by Chayes. When supplies of the sample from Westerly, Rhode Island, ran short, he tested an alternative source nearby in 1959 because the original quarry had been built over with houses.

It was common practice to plot rock compositions on a Harker diagram wherein percentages of individual oxides are plotted against SiO_2 percentages for each member of a magmatically related suite of rocks.[15] The trends were widely considered as evidence to support crystal fractionation wherein a specimen was regarded as a residue of a previous fractionation or as the starting material for subsequent fractionation.[16] In 1926 Fenner argued that the linear nature of the variation might equally well result from progressive contamination or mixing.[17] Nearly forty years later Chayes squashed the debate by pointing out the strong negative correlation between silica and other oxides irrespective of the processes involved. In short, he said the Harker diagram was of little use in discriminating between any of the processes involved in the differentiation of volcanic rocks.[18]

Almost every approach to the correlation of rocks involved plotting ratios of oxides. Chayes again applied his fundamental knowledge of mathematics to the dangers involved when ratios are plotted against each other. For example, when three variables involve two ratios and one variable is a common term, the size of the correlation will be the same whether the common term appears as the denominator, numerator, or in each. In the first two cases the sign of the correlation will be the same, but reversed in the third case. His general rule was that there are very few valid inferences to be made between any two ratios. In addition, Chayes called attention to the constraints on data presented in the percentage form by the mere act of summing the values in a "closed" array, thereby generating correlations not attributable to the original data set.[19]

The statistical approach was also applied to powder X-ray diffractometer patterns by Chayes and W. S. MacKenzie. Their concern was the experimental error in determining peak locations and the distances between peaks. They demonstrated a clear difference in results between a ground-polished surface and a powdered smear-mount. The arithmetic details were given in appendices to their paper published in 1957.[20]

As his file of published chemical analyses of rocks accumulated, Chayes recorded them first on punched cards, then magnetic tape, and eventually on the computer.[21] By 1971 he was working on the electronic storage, retrieval, and reduction of data. At first he developed a general utility sorting program with the calculations being performed at the computation center of the University of Maryland. He provided a limited number of user manuals to encourage public use of the rock information system. By 1972 he could generate bibliographies for a rock type and calculate norms, sample frequency distribution, and principal component analysis. As public interest grew, the need for incorporating nonmetric data such as geologic age, mode of occurrence, mineral assemblage, structure, texture, and alteration also grew. With the improvement of computers and programs, Chayes proposed a new base to include more of the descriptive items as well as trace elements.[22] Collecting the enormous amount of data required help from around the world, and the organizational detail was met through the International Geological Correlation Program, Project 163. A talented exchange scientist working in Chayes' laboratory, Shu Zhang Li, undertook to build a portable data management system.[23] To facilitate use of the system, Chayes organized a workshop in 1987 at Kuwait University setting up banks of minicomputers for the use of fifty scientists from thirty-seven countries. The University continued the workshop as a center of excellence, but the Gulf War terminated the project in 1990. Without a doubt, Chayes propelled the fields of statistical petrology as well as petrofabrics to international prominence.

20

NATIONAL DEFENSE CONTRIBUTIONS

World War I

Before hostilities began in 1914, it became evident that the USA would be cut off from the European sources of optical glass. Five American companies[1] undertook to make optical glass but the quality was not satisfactory by the time the USA entered the war. Because of the critical need for high quality optical glass for military fire-control instruments, methods for its manufacture on a large scale had to be developed. The Council of National Defense appealed to the Geophysical Laboratory for help because it had been engaged for many years in the study of silicate liquids, similar to optical glass, at very high temperatures. It was the only organization in the country with a staff trained in the fundamentals necessary for the manufacture of optical glass.

In April 1917 groups of staff members were sent to the various plants and assigned the responsibility for their operations, while others remained at the Laboratory to deal with specific problems. The cooperative attitude of the companies and the direct liaison with the Army through the commissioning of F. E. Wright greatly facilitated the task.

The manufacturing problems were eventually resolved by putting the secretive cookbook glass-making methods on a scientific base. Formulae were devised so that glasses of the appropriate index of refraction or other optical constants could be prepared from the necessary constituents with a minimum of trial and error. Within a short period of several weeks, Dr. Wright worked out the boundary curves for the critical three-component systems based on the published analyses of some 110 German glasses. Flint glass, for example, contains silica, lead oxide, and an alkali oxide. The actual detailed equilibria involving these end members were not published until well after the war was over. The system $K_2O-SiO_2-SiO_2$, for example, was given by F. C. Kracek and colleagues in 1929,[2] and $PbO-SiO_2$ by R. F. Geller and colleagues in 1934.[3] In addition, the relations for the commercial soda-lime-silica glasses were published by G. W. Morey and N. L. Bowen in 1925.[4]

Even the barium-rich glasses for aerial camera lenses were made on short notice. Most important contributions were made by L. H. Adams and E. D. Williamson,[5] who deduced the laws for relieving stress in glass by annealing, and by H. S. Roberts,[6] who by direct experiment was able to formulate cooling schedules for the glass pots. R. B. Sosman had investigated some of the principles governing the corrosion of the fire clays used in the pots by the molten glass.[7] Other problems such as high dispersion due to successive iron content, stones from the digestion of unsuitable clay pots, cords and striations arising from poor stirring, and strain from rolling were all investigated. Eventually some forty-two papers were published by staff members regarding optical glass. Of these, the monograph on the properties of glass by G. W. Morey, published in 1938 and revised in 1954, remains a standard reference work.[8] Although none of the twenty scientifically-trained staff had previous experience with the manufacture of glass, all used their basic knowledge of silicates to put this new US industry on a sound basis.

The expenses incurred were covered by the Carnegie Institution of Washington and no compensation was ever received for their work. The director, A. L. Day, was eventually designated as "in charge of optical glass production, War Industries Board." Day had served as a research consultant from 1905 and as vice-president of Corning Glass Works from 1918 to 1920 while on leave from the Geophysical Laboratory, and retained that title until 1936.[9] After the armistice, the records show that 95 percent of all optical glass manufactured in the USA during the war had been made under the supervision of the staff of the Laboratory.

The skills of the chemists at the Laboratory were also put to use on the fixation of nitrogen for the manufacture of explosives.[10] Experimental work on the Bucher cyanide process and the Haber process were begun in the summer of 1918, and, consequently had not proceeded far before the end of the war.

As repugnant as the task may have been, the Laboratory also investigated some of the physical constants of mustard "gas," $(C_2H_4Cl)_2S$, in response to a military request. Adams and Williamson determined the compressibility of the oily liquid, the freezing point curve to 1.8 kbar, and its latent heat of melting.[11]

World War II

The president of CIW, Vannevar Bush, helped establish the National Defense Research Committee in 1940 and served as its chairman. On June 28, 1941 Bush became director of the Office of Scientific Research and Development (OSRD), which organized and directed most of the research efforts during the war. In that summer a comprehensive program of defense research was organized, centered at the Geophysical Laboratory. Some of the staff began to collect information from military and other sources for delineating the lines of research. After the declaration of war in December, the entire staff, supplemented initially by thirty temporary employees, and all of the resources of the Laboratory, were devoted to the tasks ahead.

The director, L. H. Adams, was appointed chairman of the committee for investigating the erosion of gun barrels due to high-pressure, hot, propellant gases released on firing. As the research proceeded, studies were concentrated on the caliber-50, rapid-fire, aircraft gun where ways were sought to counteract severe swaging of the lands and thermal expansion of the barrel. The systematic studies included analysis of the corrosion products of the steel barrels and the propellant gases. By means of isotopically labeled nitrogen in the explosive charge and use of the National Bureau of Standards' mass spectrometer, tracer studies revealed the depth of penetration of the gases. Experiments were carried out in high-pressure vessels on controlled explosions to ascertain the internal ballistics and chemical products. Metal with high, hot-hardness, as well as resistance to gas erosion was inserted as short liners in barrels at the origin of rifling. The liners were then tested on a firing range on the Potomac River or in firing ranges installed under the tennis courts (later a volley ball court) behind the Upton Street Laboratory. When fired in long bursts, the existing barrels deteriorated resulting in a rapid drop in muzzle velocity, and the bullets no longer had sufficient spin for stable flight.[12] Two temporary buildings of cinder block were built especially for the testing of machine guns. One was paid for by CIW and the other by the US Government. A superior metal was found to be the cobalt-based alloy, stellite, and it became a most useful material for making hydrothermal pressure vessels after the war.

Another group was concerned with the electroplating of chromium inside the barrel after the liner, in cooperation with the electroplating group at the National Bureau of Standards. Because it was not practical to machine a taper in the large number of barrels required, a method was designed to taper the plating, with increasing thickness of the plate toward the muzzle. In this way, constriction of the bore compensated for the thermal expansion of the barrel during firing. The increased life and accuracy was documented by test firing on the Geophysical Laboratory ranges. The barrel adapted for military use contained a short stellite liner and a chrome-plated, tapered bore.

Other contributions to the war effort included the testing of mica cleavages for use in radio condensers. As in World War I, optical glass-making techniques were investigated by several staff members for fire-control instruments and search lights. Some staff members were under contract to grow large crystals of quartz, ammonium dihydrogen phosphate (ADP) and lithium sulfate hydrate ($LiSO_4 \cdot H_2O$) for plates used in submarine detecting devices.[13] Ingerson was given leave of absence to study strategic deposits of quartz in Brazil for the US Geological Survey in 1945. Goranson was also given leave of absence to conduct investigations on the hydrodynamics of shock waves in certain metals and crystals at the US Navy Santa Fe Laboratory. Work utilizing the high-pressure techniques of the Geophysical Laboratory was also carried out in the search for materials possessing special mechanical properties.

Other projects involved phosphorescent materials (Adams), rockets and special projectiles (Gibson), radar development (Morey, Ingerson), mine and
bomb disposal (Piggot), and ballistics (Kracek, Greig, Gibson).

Several of the staff members also assisted at the Department of Terrestrial
Magnetism in the development of the proximity fuse for artillery shells. That
device was considered to be "the most important technical improvement in
weaponry to come out of World War II."[14]

Almost five years in the life of the Geophysical Laboratory were devoted
to war work. The regular staff was paid by CIW; however, the costs of the
temporary employees and extra expenses were provided by the government. In
1946 the war work was phased out, the reports written, and a comprehensive
review undertaken by Adams of the scientific programs in the light of the
irreversible changes brought about by World War II.[15]

World War III (?)

The world is now under the threats of both bioterrorism and chemical terror-
ism in addition to the potential devastation of a nuclear event. The biological
threat is of special concern to Laboratory staff who have been working on
a device that will detect hundreds of separate molecules. The instrument
currently employs all commercially available technology and relies on antibody arrays.[16] Each antibody is specific for a molecule indicating either viable
or fossil contaminating organisms as well as prebiotic chemicals and can be
printed into a high-density array format so that thousands of reactions can be
monitored at the same time. This instrument called MASSE is being developed in collaboration with both US and European collaborators including
NASA, Lockheed Martin, Centro de Astrobiologia, Montana State University, and Oceaneering Space Systems. This technology may be tailored for
a number of other applications including medical as well as biological war-
fare considerations by changing the choice of antibodies. In addition, it is
intended that the instrument will be used to detect life on other planets.

21

PUBLICATIONS

In the presentation of the original Trust Deed on January 28, 1902, Mr. Carnegie specified one of the aims of the Carnegie Institution of Washington as being "To ensure the prompt publication and distribution of the results of scientific investigation" (*CIW Year Book* 1, p. xiv). The entire array of publication media was used to accomplish this end. Initially, the Laboratory reported its results in the "Annual Report of the Director" published in the *CIW Year Books*. The reports were aimed at a scientific audience and provided detailed, technical information on current research activities. Projects were usually brought to a stage of fruition before publishing in the appropriate journals.

With *Year Book* 83 (1983–84), the detailed Annual Report was terminated by the president of CIW and was replaced with a smaller narrative description written for the general reader. The work was indeed accessible to a broader audience at less cost, but requests from many scientific colleagues for a revival of the more informative old-style Report led in 1997–98 to a new "Annual Report of the Director" published entirely within the Geophysical Laboratory using desktop-publishing techniques. Four such Reports were issued and distributed to 3000 researchers and institutions in the USA and abroad before succumbing to fiscal pressures in 1992. Today, essays by the director in each *Year Book* continue to provide descriptions of current areas of investigation. Comprehensive "Indices of the Annual Reports of the Director, Geophysical Laboratory" for 1905–1980 were published on the occasion of the 75th anniversary of the Laboratory. No update has been compiled, but most of the scientific content in subsequent Reports is indexed in the GeoRef database produced by the American Geological Institute.

Journals

Over 3100 papers had been published by Laboratory staff by the end of 2002 according to the numbered list of publications (Figure 21.1). The large drop during the 1940s was no doubt due to confidential and secret war work (Chapter 20). The spikes in 1920 and 1959 were due, respectively, to an array of papers on optical glass accumulated during World War I and the volume "Researches in Geochemistry" organized by P. H. Abelson. The spikes in

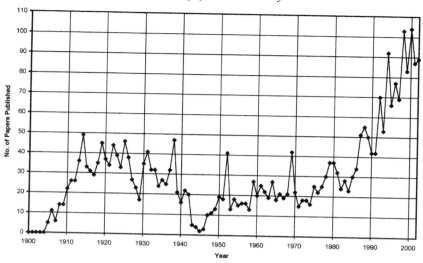

Figure 21.1   Number of papers published per year by Laboratory staff in formal journals and books since inception in 1905 to present.

1938, 1952, and 1969 can be attributed to volumes in honor of A. L. Day, N. L. Bowen, and J. F. Schairer, respectively. The rapid increase beginning in the 1980s is presumably due to the demand by funding agencies for rapid publication of results in short papers. Eventually those results were collated and analyzed in review papers or books. Support for this view may be found in the rapid growth of external funding for the Geophysical Laboratory. In Figure 21.2 the total operating expenses for 1982–83[1] to the present are displayed and the amount of Carnegie Funds provided. It is evident that the drop in the ratio of Carnegie funding in 2001–2002 to 41 percent of the total funding raises serious issues in regard to external impositions on the research programs and their in-depth pursuit.

In addition to the formal numbered publications such as journal papers and books, there are some 250 miscellaneous publications such as extended abstracts, short biographies, commentaries and discussions, obituaries, memorials, presentation and acceptance of awards, book reviews, committee reports, laboratory procedures, handbook and encyclopedia contributions, historical reflections, and occasional data tables.

For many years, announcements of new publications by Laboratory personnel were distributed roughly semi-annually to a large mailing list, which by the late 1980s had grown to around 3000 individuals, libraries, and research institutes worldwide. These announcements typically included a postcard-type return form, by which reprints of specific papers could be requested. Large quantities of reprints (in some cases, 1000 copies or more) were kept

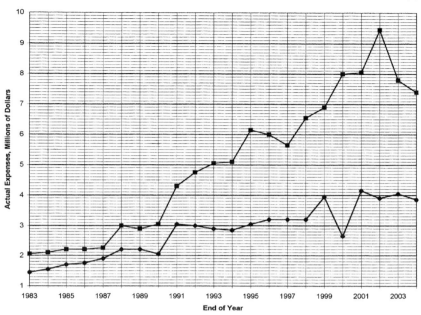

Figure 21.2 Upper graph: actual total expenses ($ smillion) for the GL from 1983 to the present. Lower graph: funding provided by CIW. The difference in dollars between total expenses and Carnegie funding was met by federal and private funds.

on hand to fill anticipated demand. The reprints, like the Annual Report of the Director, were always distributed free of charge. With declining demand, due to the widespread availability of inexpensive photocopying, and sharp reductions in the operating budget, the announcement card system was discontinued in 1992, with the mailing of "Current List No. 182" being the last. Since that time, bibliographies of Laboratory publications continue to be published annually in the *Carnegie Institution Year Book*. Since the late 1990s, online versions of the bibliographies are also posted on the Laboratory website. Current demand is comparatively light, with most requests now arriving via email, but the library continues to furnish print copies of recent publications at no charge. The transition to online publishing of scientific journals has heightened interest in distributing GL papers in electronic form, but that goal has remained elusive, owing to the complex copyright restrictions in the digital arena.[2]

The collected works of a number of staff members have been assembled for reference use on-site. The first electronic collection of papers was that of H. S. Yoder, Jr., 1950–2000; it was issued by the Laboratory on a CD-ROM in 2001. Bibliographies are maintained in the library for all investigators and are available for reference use. Partial bibliographies often accompany award presentations and full bibliographies may be found in the Memorials such as those of the National Academy of Sciences.[3]

Monographs

In addition to the *Year Books*, the CIW has published a series of monographs that now exceeds 600 volumes. Reference has been made to some of them in the foregoing chapters especially those on volcanology and field petrology. A catalogue is available from CIW.

Patents

On very rare occasions the CIW patented work done at the Laboratory. Although all work was to be freely available to the public, some discoveries had considerable economic potential, which, if patented by others, would restrict their use. Among those granted are the following:

US 4,290,893, September 22, 1981 "Separation of amino acids"
Inventors: E. P. Hare and E. Gil-Av

US 4,339,252, July 13, 1982 "Apparatus for producing solid hydrogen"
Inventors: P. M. Bell and H-k. Mao

US 4,383,950, June 7, 1983 "Methods and apparatus for producing solid hydrogen at high pressure"
Inventors: P. M. Bell and H-k. Mao

US 5,981,094, November 9, 1999 "Low compressibility carbon nitrides"
Inventors: D. Teter and R. J. Hemley

US 6,543,295, April 8, 2003 "High pressure anvil and optical window"
Inventors: J. Xu and H-k. Mao

Archives

Approximately 140 linear feet of records are preserved in the archives. These include administrative records, fiscal and personnel records, architectural plans, and field and laboratory notebooks. Small collections of the personal correspondence of N. L. Bowen, F. A Perret, and other luminaries provide some historical insight. A collection of over 800 historic photographs document the development of instrumentation in the Laboratory. The special investigations such as the drilling in the geyser basin at Yellowstone National Park are recorded. In general, the correspondence of the directors has not been retained, but the files of investigators retain much pertinent historical information. Plans are underway to improve the organization, preservation, and accessibility of the archives. Finding aids are high on the list of initiatives.

Citation indices

Lists of most-cited papers from the earth sciences journals have revealed some measure of the quality and impact of the papers from the Geophysical Laboratory. Although the number of citations is not necessarily a valid indicator of quality, they do measure in a crude way the impact of the work in a research field. For example, in the period 1945–54, seven papers by Laboratory staff are listed among the twenty most cited.[4] The authors were N. L. Bowen, O. F. Tuttle, J. F. Schairer, J. W. Greig, and S. B. Hendricks. The value of using quantitative physicochemical methods to study geological problems was clearly recognized, especially in applications to the granite problem.

Among the one hundred most-cited articles in the physical sciences for the 1961–1982 period,[5] the 1962 paper by Yoder and Tilley is listed.[6] It is also the most cited paper in the *Journal of Petrology* during its run since 1959. Only three other papers in geophysics (geomagnetic reversals, seismology, and sea-floor spreading) were among the most cited. The CIW was cited by Essential Science Indicators, an ISI/Thomson Scientific web-based evaluation tool, as eighth among geoscience research institutions worldwide for citation impact, i.e., citations per paper in the 1991–2001 period.[7]

Lectures

The work of the Geophysical Laboratory has also been disseminated by staff through lectures and courses at universities, symposia, and conferences. Staff have frequently been invited to give lectures on specific topics, to present a series of lectures involving the historical development of an idea, or to generate an entire course extending through a term at a university. Conferences and symposia have been particularly rewarding because the questions and discussions initiate new research directions or improvements in techniques. A specific dividend of these venues has been the excitement of interest in the work of the Laboratory for potential pre- or postdoctoral fellowship candidates. The enthusiasm of the staff in presenting their work has been a large factor in attracting student candidates, and encouraging them to take the prerequisite courses at their university for experimental work.

Seminars

Seminars have been regularly scheduled twice a week except during the month of August and on government holidays. The principal benefit of such seminars arises from the discussion of new data and concepts. Candidates for fellowship, in particular, have had an opportunity to display their knowledge and enthusiasm for a subject of their choice.

Informal discussion sessions were initiated by A. L. Day, the first director, at a community lunch table. He sat the head of the table, according to N. L. Bowen, and encouraged all to participate. That format continued until the early 1970s when the lunch room was converted to a laboratory for a new staff member. Since 1990 some of the Laboratory staff have joined the DTM lunch club and participated in a wider range of discussions there.

As participants in the NASA Astrobiology Institute, some staff take part in the new format of video conferencing. In this way large numbers of small groups in the Astrobiology program at diverse locations can exchange views face to face without the expenses incurred in travel and, especially, in loss of research time. A very large video screen, cameras, and microphones provide coverage of the event at least once a month.

22

SUPPORT STAFF

The success of the Geophysical Laboratory has been due in no small measure to the talented support staff. Especially skilled technicians manufactured the new types of apparatus, built the electrical controls, assembled the trains of glassware for chemical processing, and erected the frames for equipment assemblies. In addition, power and heat generation was critical in the early days, as were building and grounds maintenance and laboratory room reconstruction and additions. To these efforts were added the stockroom, mailroom, library, accounting, and secretarial services for correspondence, manuscript preparation, and telephone. Needless to say, with a small staff each person was required to fulfill a range of needs.

Mechanical shop

Almost every experiment undertaken in the new fields of experimental geophysics and geochemistry required new devices. A shop was fitted with lathes, mills, drill presses, welding and grinding equipment, and the wide array of hand tools required. Originally, each machine was driven by a belt turned by a pulley near the ceiling. These were replaced by individual electric motors for more versatile operation after World War II. A very large lathe and a milling machine were acquired by purchase from military surplus. The number of instrument makers varied from several to a dozen during World War I. Each had special skills for handling hard steels[1] and the softer metals such as brass and copper. Where one would specialize in cutting brass for optical equipment, another would build the furniture or supports out of wood. (The wood shop was a particular favorite of the younger staff members who used the equipment on weekends to build their home furniture.) Another mechanic specialized in cutting rocks and making thin sections for optical study. A darkroom was provided for special optical studies and for photographic work. The large glass plate camera carried to Yellowstone National Park was installed on the frame for copy work. Another specialist was required to build the devices for controlling the temperature of experimental furnaces.

The mechanical shop was under the direction of a master mechanic or foreman (Table 22.1). Initially, drawings prepared by the scientific staff were handed to the foreman who then assigned the project to the mechanic most

Table 22.1 *Shop leadership*

Name	Title	Service dates
Benjamin D. Chamberlin	Master Mechanic	1904–1910
Charles M. Shaw	Master Mechanic	1910–1930
John Jost	Chief Mechanic	1931–1945
Francis A. Rowe	Chief Mechanician	1946–1971
Wilbur Roos	Acting Shop Foreman	1975–1976
Charlie A. Batten	Shop Foreman	1977–1986
Andrew J. Antoszyk	Shop Foreman	1987–1999
Stephen D. Coley	Shop Foreman	2000–

familiar with the materials involved. After World War II, a scientific staff
member could go directly to a mechanic and discuss his needs. This arrangement was preferred in that the ideas and suggestions of the mechanic could
be incorporated into an improved design. Although many pieces of apparatus
were successful because of the contributions of the mechanics, their names
were rarely included in a paper, except perhaps as an acknowledgment. One
cannot over-emphasize the significant contributions the mechanics made in
the construction of successful apparatus.

Power generation

The initial mode of power generation was by coal-fired steam boilers; the
exhaust steam from the power plant was utilized to heat the entire building
through forced air circulation. The constant direct current needed for the
many platinum-wound, high-temperature furnaces was supplied by five large
acid battery cells maintained in a separate basement room (see Figure 1.5).
Compressed air, illuminating gas, and vacuum lines were provided by individual pipes to those laboratories having exhaust hoods. In 1931 a new power
plant was constructed to the west of the main building that used diesel oil. It
is recalled that a 4000-gallon fuel storage tank was installed next to the X-ray
building, behind which was a shed for storage of acids, volatile chemicals,
and other hazardous materials. The flat roof of the new power plant served
as a parking lot accessible to the side door of the Laboratory.

Electronics shop

Displayed in Figures 2.3, 2.4, and 2.5 are some of the temperature regulating
devices built at the Laboratory by shop personnel, including the electrician
Mr. E. C. Huffaker and a physicist Mr. J. L. England. In July 1965, an elec-
tronics technician, C. G. Hadidiacos, was hired primarily to maintain the

new electron microprobe. His new designs for the instrument resulted in having the data recorded on punch cards that could be processed by a computer. The chief advantage was that 20–50 analyses could be made in a day. By 1970 he and L. W. Finger had automated the microprobe so that the computer controlled all the collection process and data reduction. With his new technique a complete analysis could be produced every two to three minutes. Early on, Hadidiacos replaced many of the temperature controllers using a variac with solid-state temperature controllers. In a cooperative effort, Finger and Hadidiacos automated the single-crystal, X-ray diffractometer in 1972 and provided a programmable temperature controller for the diffractometer system. The increasing demand for electronic maintenance led to the hiring of another technician, David George, in 1975. New designs emerged that gave full control of the temperature regulator systems and maintained a log of the temperature profile during an experiment. With the purchase of an IBM personal computer, they designed an interface board so that up to four furnaces could be controlled. That design has been adapted to high-pressure equipment and is currently used today.

Word processing and computers

The devoted and skillful secretaries at the Laboratory have relieved the staff of many tasks. The preparation of letters and manuscripts on mechanical and eventually electronic typewriters is now carried out on word processors either by the staff member directly or occasionally by a secretary. The email system would appear to relieve the time previously used for paper mail, but the volume has increased so dramatically that the staff are not always able to respond in reasonable times. The speed of email has been beneficial in obtaining discussion of new ideas while in the development stages. More than 200 desktop computers are now available for use by the staff, in addition to the computers devoted to the apparatus.

In the 1960s the bulk of the computing was done using equipment available at the Van Ness Campus of the National Bureau of Standards (NBS, now National Institute of Standards and Technology). The GL X-ray crystallographer, Charles Burnham, also had a contract with Control Data Corporation in Rockville, Maryland, where paper tapes could be converted to magnetic tapes for processing at the Bureau of Standards. When the NBS moved to Gaithersburg, Maryland, new arrangements had to be made. Initially a telephone connection was made to IBM to process the punch card information. That proved to be too slow and limited, so a remote-job-entry terminal was leased from Univac. They informed L. Finger in 1969 of the availability of a second-hand computer from the American Bakers and Confectioners Union, which became the first computer owned by the Laboratory. It was connected by telephone lines to the University of Maryland. The magnetic

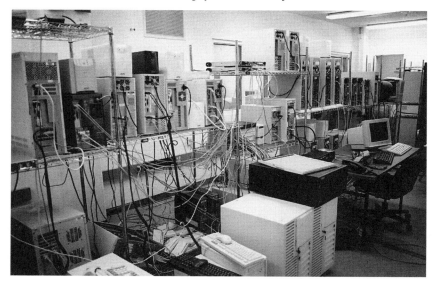

Figure 22.1   Computer room servicing both the GL and DTM (Archives, GL).

tapes were sent by courier, but the printout was retrieved from the terminal. The unit was programmable with a plug board and there were programs for off-line listing, card-deck copying, and provision for reprint mailing labels. In 1971 the first new minicomputer was purchased on a NASA grant: a PDP-11/20 from the Digital Equipment Corporation. Through the skills of Finger and Hadidiacos the system was used for both the microprobe and the X-ray diffractometer. A second PDP-11 was purchased in 1977 to provide for in-house capabilities. With upgrades and enlargement for multiple users, it hosted the first use of desktop computing at the Laboratory. By 1981 the PDP-11 had been replaced with the VAX 11/780 system, and it was then possible to put a terminal on the desk of anyone who wanted one (see Figure 22.1). That system was in turn replaced by a VAX 3300 in 1989, which was replaced in 1995 by a Cray J916/8-1024 and upgraded in 1999. In 1992 the first HP-UNIX series 400 was purchased to run and control the JSM 8900 Microprobe. It was replaced in 1994 by series 700i RISC computer and is presently used to run the microprobe.

Office management

The administration of the daily functions of the Laboratory were generally carried out by the director's secretary. The first stenographer, later secretary, on record was Alice F. Susan who joined on January 1, 1907 and resigned on November 30, 1922. Apparently A. L. Day met Ruth S. Easling while working at Corning Glass Works and persuaded her to serve as secretary from August 1, 1921. When Day's wife returned to her native Germany with three of their children, divorce ensued, and Day eventually married Easling in 1933, who then resigned from the Laboratory. Presumably most of the administrative duties were taken over by the Accountant, J. Harper Snapp. Some of these duties were acquired by Mildred O. Giblin on July 31, 1944 under the directorship of L. H. Adams. She served until August 31, 1950. Adams petitioned the president of CIW, Vannevar Bush, for an administrative assistant and E. G. Fullinwider was hired as Executive Officer on July 5, 1950. Fullinwider was a decorated retired Rear Admiral from the US Navy who was described as having the "right combination of tact and firmness" (L. H. Adams, letter of July 3, 1950 to V. Bush). No doubt due to his military training, Admiral Fullinwider prepared a detailed book for the operation, organization, maintenance, and procedures to be used by the staff. (The book was similar to the Watch-Quarter-Station Bill used aboard naval ships for assignment of duties and space.) With the retirement of Adams in 1952, G. W. Morey served as acting director. On the appointment of P. H. Abelson as director, V. Bush wrote to Fullinwider that "with a younger man coming into the office the period of need for a senior man of your capability is drawing to a close. . . ." (Bush letter of March 5, 1953), and he resigned August 31, 1953. Abelson hired A. David Singer, whom he had known while working at DTM. Singer was a very able administrator and even found time to handle Abelson's correspondence relating to his editorship of the *Transactions of the American Geophysical Union*. Singer's service extended through most of the directorship of H. S. Yoder, Jr. An illness briefly prevented his working full time and Singer died of cancer on 13 March 1985. Fortunately, Marjorie E. Imlay, an experienced stenographer, was on the staff and very quickly took on the duties of the director's assistant. She served through Yoder's term as director, and continues to serve as executive assistant to the present director, Wesley T. Huntress, Jr.

23

FUTURE OPPORTUNITIES

The preceding chapters have sampled the array of scientific problems undertaken at the Geophysical Laboratory. Successful investigations always reveal more critical and pertinent problems to be resolved. With the complete freedom of the individual investigator to choose these problems, it is clear that they will propel the Laboratory into an exciting and productive future. Every field of research has obvious potentials that can be recognized now, but it is unlikely that they will be restrained by new discoveries.

Although only five-component systems have been examined in experimental igneous petrology, the need to address the natural ten-component system is great, but hindered by the lack of a method for illustrating and examining the physicochemical changes. If the philosophy of proceeding with one additional component at a time is adopted, eventually a plan will be devised for appreciating the laboratory results and relating them to the natural rocks. Because the entire range of pressure and temperature within the Earth can be sustained in the laboratory, the field is open to examine any theory regarding the structure and composition of the Earth. The tools are available for characterizing liquids within the Earth in the presence or absence of volatiles. The crystals produced can be characterized no matter how small, both in composition and in structure. The criteria are being defined for ascertaining temperature and pressure within the entire Earth from the composition of coexisting crystals.

A field of research that should be of great concern to the nation is ore geochemistry. The current lack of interest in mineral resources essential to the economy is likely to undergo significant change as issues of accessibility to limited resources come to the fore over the next decade. The techniques for resource discovery will require refinement as well as a focus on more detailed geological mapping.

There is little doubt that the fastest growing scientific field will be in biogeochemistry. The recent discoveries have opened more new directions than can be pursued at the Geophysical Laboratory, particularly those concerning the origins of life. The development of the protein chip, aside from its use in the exploration of other planets, will be useful in combating the bioterrorism threat on Earth and may become the primary application. The molecular and isotopic signatures that indicate origin and alterations of biochemicals will be an important aspect.

These exciting opportunities will no doubt be dwarfed by the new ventures generated by a very talented and interactive staff. The tradition of creating a vital and existing scientific future for the Laboratory seems assured. The Carnegie Institution has a long history of finding new, emerging scientific fields. It is this tradition that will create a vital and exciting scientific future for the Laboratory. Independence as a privately endowed research institution, and the freedom to pursue risky but potentially high-payoff new ventures in science, are the enduring characteristics that will propel the Geophysical Laboratory into an exciting, productive, and respected future in science.

NOTES

1 Introduction

1. J. Fröbel, "Entwurf eines Systems der geographischen Wissenschaften I.," *Mittheilungen aus dem Gebiete der Teoretischen Erdkunde* 1 (Fussli, Zurich, 1834), 1–35.
2. Translation from the original German was kindly provided by Dr. Louis Brown.
3. M. C. Rabbitt, *Minerals, Lands, and Geology for the Common Defense and General Welfare*, 3 volumes (US Geological Survey, US Government Printing Office, 1979–1986).
4. C. Barus and V. Strouhal, "The electrical and magnetic properties of the iron carburets," *US Geol. Surv. Bull.* 14 (1885).
5. E. L. Yochelson, "Andrew Carnegie and Charles Doolittle Walcott: The origin and early years of the Carnegie Institution of Washington." In G. A. Good (ed.), *The Earth, the Heavens and the Carnegie Institution of Washington* (American Geophysical Union, 1994), pp. 1–19.
6. *Ibid.*, p. 5.
7. The first phase of the new twelve-storey building, one of the first steel structures in Washington, was constructed on the original site at 1401 Pennsylvania Avenue and continued to operate until 1968. After restoration and the construction of a new office building and retail shops, it reopened as the Willard Inter-Continental Washington. It is described as "The Residence of the Presidents" (historical brochure) because it has hosted every president from Franklin Pierce in 1853 to the present. It served as a hostelry in some form since 1816, but did not become a major force in the political and social life of Washington until 1850.
8. C. K. Wead, "Philosophical Society of Washington," *Science* 13 (1901), 190.
9. Library of Congress, Becker Papers, Box 26.
10. *Ibid.*
11. G. F Becker, "Project for a Geophysical Laboratory: Appendix 1 to Report of Advisory Committee on Geophysics," *CIW Year Book* 1 (1903), 44–58.
12. R. S. Woodward, "Report of the President of the Institution," *CIW Year Book* 4 (1906), 17–33, at p. 32.
13. There is no record that the bibliography for geophysics was ever published by CIW. The data were presumably published by F. B. Weeks as part of his inclusive bibliography and index of North American geology: "Bibliography and index of North American geology, paleontology, petrology, and mineralogy for the years 1901–1905," *US Geol. Surv. Bull.* 301 (1906).
14. E. L. Yochelson and H. S. Yoder, Jr., "Founding the Geophysical Laboratory, 1901–1905: A scientific bonanza from perception and persistence," *Geol. Soc. Am. Bull.* 106 (1994), 338–350.
15. C. R. Van Hise, "Report on Geophysics," *CIW Year Book* 2 (1904), 173–184.

16. W. Cross, J. P. Iddings, L. V. Pirsson, and H. S. Washington, "A quantitative chemico-mineralogical classification and nomenclature of igneous rocks," *J. Geol.* 10 (1902), 555–690.

17. H. S. Yoder, Jr., "Development and promotion of the initial scientific program for the Geophysical Laboratory." In Good (ed.), *The Earth, the Heavens*, pp. 21–28.

18. According to the minutes of the third meeting of the Board of Trustees, they met in Room 1001 of the New Willard Hotel. Mr. Carnegie was present in the morning session but is not listed as present after the 1–2 p.m. recess when the funding of the Geophysical Laboratory, to be built in Washington, was passed.

19. H. W. Servos, "To explore the borderland: the foundation of the Geophysical Laboratory of the Carnegie Institution of Washington," *Hist. Stud. Phys. Sci.* 14 (1983), 147–185, at pp. 174–175.

20. A. F. Williams, *The Genesis of the Diamond* (Ernest Benn, Ltd, London, 1932), 2 volumes.

21. A. L. Day, E. T. Allen, and J. P. Iddings, *The Isomorphism and Thermal Properties of the Feldspars* (Carnegie Institution of Washington Publication 31, 1905).

22. *Ibid.*

23. D. Cahan, *An Institute for an Empire: The Physikalische-Technische Reichsanstalt, 1871–1918* (Cambridge University Press, Cambridge, 1989).

24. A topographic map for 1859 shows the site to be an open field with wooded areas surrounding the site. The 1899 map showed in addition the farmhouse called "Rock of Dunbarton" to the west of the site. A half-acre of land was given by the Institution to the District of Columbia for construction of Upton Street, south of the Laboratory.

25. *Carnegie Institution of Washington (CIW) Year Book* 5 (1907), 23.

26. *CIW Year Book* 1 (1903), 44–58.

27. *CIW Year Book* 2 (1904), 185–194.

28. L. Mechlin, "The work of Wood, Donn, and Deming," *Archit. Rec.* 19 (1906), 248–251. An innovative structural feature of the building – to provide for thermal insulation during Washington's hot and humid summers – was the construction of an inner wall of ordinary brick protected from the heat by a 6-inch insulating layer of hollow terracotta with horizontal air spaces closed at the ends.

29. D. Hoffmann, *Gustav Magnus und sein Haus* (Verlag für Geschichte der Naturwissenschaften und der Technik, Stuttgart, 1995), p. 119.

30. H. S. Yoder, Jr., "Scientific highlights of the Geophysical Laboratory, 1905–1989," *Annual Report of the Director, Geophysical Laboratory, 1988–1989* (1989), p. 146.

31. The title of "Earth and Planetary Science Department" was suggested later by GL staff members. The Joint Feasibility Study Committee Report (p. 25) provided the common title of "The Geophysical Laboratories of the Carnegie Institution of Washington."

32. *CIW Year Book* 84 (1985), 3.

33. *CIW Year Book* 85 (1986), 4.

34. F. Press, quoted in "Groundbreaking ceremony held for new research building," *CIW Newsletter*, March (1989), 1–3.

35. The renovated building was named the Geron P. Johnson Hall in memory of the late infant son of Sheila and Robert Johnson who made a $1 million gift to the Levine fundraising campaign. Mr. Johnson was founder and chairman of the Black Entertainment Television Network and his wife was executive vice-president of

BET and a Trustee and Campaign Steering Committee member of the Levine School of Music.

2 Igneous petrology

1. R. B. Sosman, "Temperature scales and silicate research," *Am. J. Sci.*, Bowen Volume 250A (1952), 517–528.
2. A. L. Day, E. T. Allen, and J. P. Iddings, *The Isomorphism and Thermal Properties of the Feldspars* (Carnegie Institution of Washington Publication 31, 1905).
3. E. S. Shepherd, G. A. Rankin, and F. E. Wright, "The binary systems of alumina with silica, lime, and magnesia," *Am. J. Sci.*, 4th series, 28 (1909), 293–333.
4. N. L. Bowen, "The melting phenomena of the plagioclase feldspars," *Am. J. Sci.*, 4th series, 35 (1913), 577–599.
5. J. F. Schairer also determined the albite–anorthite system using the quenching technique: R. R. Franco and J. F. Schairer, "Liquidus temperatures in mixtures of the feldspars of soda, potash, and lime," *J. Geol.* 59 (1951), 259–267; J. F. Schairer, "Melting reactions of the common rock-forming oxides," *J. Am. Ceram. Soc.* 40 (1957), 215–235.
6. G. W. Morey, "A comparison of the heating-curve and quenching methods of melting-point determinations," *J. Washington Acad. Sci.* 13 (1923), 326–329.
7. A. L. Day, R. B. Sosman, and E. T. Allen, *High Temperature Gas Thermometry* (Carnegie Institution of Washington Publication 157, 1911), p. 129.
8. A. L. Day and R. B. Sosman, "The nitrogen thermometer scale from 300° to 630° with a direct determination of the boiling point of sulphur," *Am. J. Sci.* 4th series, 33 (1912), 517–533.
9. L. H. Adams, "Calibration tables for copper-constantan and platinum-platinrhodium thermoelements," *J. Am. Chem. Soc.* 36 (1914), 65–72.
10. The fixed boiling and melting points below gold are listed in Sosman, "Temperature scales and silicate research," p. 522.
11. W. P. White and L. H. Adams, "A furnace temperature regulator," *Phys. Rev.* 14 (1919), 44–48.
12. H. S. Roberts, "A furnace temperature regulator," *J. Wash. Acad. Sci.* 11 (1921), 401–409.
13. H. W. Roberts, "The Geophysical Laboratory furnace thermostat," *J. Opt. Soc. Am.* 11 (1925), 171–186.
14. C. Hadidiacos, "Solid-state temperature controller," *J. Geol.* 77 (1969), 365–367.
15. C. Hadidiacos, "Temperature controller for high-pressure apparatus," *CIW Year Book* 71 (1972), 620–622.
16. C. Hadidiacos, "Linear programmable temperature controller," *CIW Year Book* 76 (1977), 664–668.
17. N. L. Bowen and J. W. Greig, "The system: Al_2O_3-SiO_2," *J. Am. Ceram. Soc.* 7 (1924), 238–254; J. H. Welch, "A new interpretation of the mullite problem," *Nature* 186 (1960), 545–546, fig. 2.
18. G. A. Rankin and F. E. Wright, "The ternary system CaO-Al_2O_3-SiO_2," *Am. J. Sci.*, 4th Series, 39 (1915), 1–79; E. F. Osborn and A. Muan, *Phase Equilibrium Diagrams of Oxide Systems* (American Ceramic Society, 1960), plate 1.
19. G. W. Morey and N. L. Bowen, "The ternary system sodium metasilicate–calcium metasilicate–silica," *J. Soc. Glass Tech.* 9 (1925), 226–264.

20. L. H. Adams, "List of systems investigated at Geophysical Laboratory," *Am. J. Sci.*, Bowen Volume 250A (1952), 1–26.

21. R. M. Hazen and M. H. Hazen, *Indices of the Annual Reports of the Director of the Geophysical Laboratory, 1905–1980* (Geophysical Laboratory, Carnegie Institution of Washington, 1981).

22. N. L. Bowen, "The binary system: $Na_2Al_2Si_2O_8$ (nephelite, carnegieite) – $CaAl_2Si_2O_8$ (anorthite)," *Am. J. Sci.* 4th series, 33 (1912), 551–573.

23. *Ibid.*

24. Bowen, "Melting phenomena."

25. N. L. Bowen, "The ternary system: diopside-forsterite-silica," *Am. J. Sci.*, 4th series, 38 (1914), 207–264.

26. O. Anderson, "The system anorthite-forsterite-silica," *Am. J. Sci*, 4th series, 39 (1915), 407–454.

27. N. L. Bowen, "The crystallization of haplobasaltic, haplodioritic and related magmas," *Am. J. Sci.*, 4th series, 40 (1915), 161–185.

28. Rankin and Wright, "The ternary system $CaO-Al2O_3-SiO_2$."

29. N. L. Bowen, "The reaction principle in petrogenesis," *J. Geol.* 30 (1922), 177–198.

30. G. A. Rankin and H. E. Merwin, "The ternary system $CaO-Al_2O_3-MgO$," *J. Am. Chem. Soc.* 38 (1916), 568–583.

31. G. A. Rankin and H. E. Merwin, "The ternary system $MgO-Al_2O_3-SiO_2$," *Am. J. Sci.*, 4th series, 45 (1918), 301–325.

32. J. B. Ferguson and H. E. Merwin, "The ternary system $CaO-MgO-SiO_2$," *Am. J. Sci.*, 4th series, 48 (1919), 81–123.

33. J. B. Ferguson and A. F. Buddington, "The binary system akermanite-gehlenite," *Am. J. Sci.*, 4th series, 50 (1920), 131–140.

34. J. W. Greig, "Liquid immiscibility in silicate melts," *Am. J. Sci.*, 5th series, 13 (1927), 1–44, 133–154.

35. N. L. Bowen, *The Evolution of the Igneous Rocks* (Princeton University Press, 1928), 334 pp.

36. G. F. Becker, "Fractional crystallization of rocks," *Am. J. Sci.*, 4th series, 4 (1897), 257–261.

37. H. S. Yoder, Jr., *The Evolution of the Igneous Rocks: Fiftieth Anniversary Perspectives* (Princeton University Press, 1979), 588 pp.

38. J. F. Schairer and N. L. Bowen, "Preliminary report on equilibrium relations between feldspathoids, alkali-feldspars, and silica," *Trans. Am. Geophys. Union*, *16th Annual Meeting* (National Research Council, Washington, DC, 1935), pp. 325–328.

39. N. L. Bowen, "Recent high-temperature research on silicates and its significance in igneous geology," *Am. J. Sci.*, 5th series, 33 (1937), 1–21.

40. J. F. Schairer, "The system $CaO-FeO-Al_2O_3-SiO_2$: I. Results of quenching experiments on five joins," *J. Am. Ceram. Soc.* 25 (1942), 241–274.

41. J. F. Schairer and N. Morimoto, "The system forsterite-diopside-silica-albite," *CIW Year Book* 58 (1959), 113–118; J. F. Schairer and H. S. Yoder, Jr., "The nature of residual liquids from crystallization, with data on the system nepheline-diopside-silica," *Am. J. Sci.*, Bradley Volume 258-A (1960), 273–283; J. F. Schairer and H. S. Yoder, Jr., "Crystallization in the system nepheline-forsterite-silica at 1 atmosphere pressure," *CIW Year Book* 60 (1961), 141–144; I. Kushiro, "Determination of liquidus relations

in synthetic silicate systems with electron probe analysis: the system forsterite-diopside-silica at 1 atmosphere," *Am. Mineral.* 57 (1972), 1260–1271; H. S. Yoder, Jr., *Generation of Basaltic Magma* (National Academy of Sciences, Washington, DC, 1976).

42. H. S. Yoder, Jr., *Generation of Basaltic Magma* (National Academy of Sciences, Washington, DC, 1976).

43. *Ibid.*

44. H. S. Yoder, Jr. and C. E. Tilley, "Origin of basalt magmas: an experimental study of natural and synthetic rock systems," *J. Petrol.* 3 (1962), 342–532.

45. D. C. Presnall, "An algebraic method for determining equilibrum crystallization and fusion paths in multicomponent systems," *Am. Mineral.* 71 (1986), 1061–1070.

46. J. W. Greig, E. S. Shepherd, and H. E. Merwin, "Melting temperatures of granite and basalt," *CIW Year Book* 30 (1931), 75–78; Yoder and Tilley, "Origin of basalt magmas."

47. R. B. Sosman and J. C. Hostetter, "The reduction of iron oxides by platinum, with a note on the magnetic susceptibility of iron-bearing platinum," *J. Wash. Acad. Sci.* 5 (1915), 293–303.

48. N. L. Bowen and J. F. Schairer, "The system FeO-SiO_2," *Am. J. Sci,* 5th series, 24 (1932), 177–213.

49. Schairer, "The system CaO-FeO-Al_2O_3-SiO_2."

50. H. G. Huckenholz and H. S. Yoder, Jr., "Andradite stability relations in the $CaSiO_3$-Fe_2O_3 join up to 30kb," *Neues Jahrb. Miner. Abh.* 114 (1971), 246–280.

51. L. S. Darken and R. W. Gurry, "The system iron-oxygen. I. The wüstite field and related equilibria," *J. Am. Chem. Soc.* 67 (1945), 1398–1412.

52. D. H. Lindsley, "Fe-Ti oxides in rocks as thermometers and oxygen barometers," *CIW Year Book* 62 (1963), 60–66.

53. F. E. Wright, "A new petrographic microscope," *Am. J. Sci.,* 4th series, 28 (1910), 407–414.

54. H. E. Merwin and E. S. Larsen, "Mixtures of amorphous sulphur and selenium as immersion media for the determination of high refractive indices with the microscope," *Am. J. Sci.,* 4th series, 34 (1912), 42–47.

55. F. E. Wright, "A new goniometer lamp. A containing device for salts used as sources for monochromatic light," *Am. J. Sci.,* 4th series, 27 (1909), 194–195.

56. P. Debye and P. Scherrer, "Interferenzen an regellos orientierten Teilchen im Röntgenlicht," *Physik. Zeit. I.* 17 (1916), 277–283. The technique appears to have been developed independently by A. W. Hull in the USA at the General Electric Company. His other inventions include the thyratron tube, a gas-filled tube used to control high-temperature furnaces, for example. A. W. Hull, "A new method of X-ray crystal analysis," *Phys. Rev.* 10 (1917), 661–696.

57. The director, Dr. L. H. Adams, was so enamored of the machine and its calibration that he would not let anyone use it for months. A staff member briefly placed a piece of uraninite near the Geiger counter to introduce a spurious peak in the pattern of diamond that the director was using for calibration. Unable to account for the event, the director became discouraged and eventually released the machine for staff use.

58. J. Van den Heurk, "Norelco powder X-ray diffraction specimen changer," *CIW Year Book* 53 (1954), 140–141.

3 Pressure

1. C. Barus, "High-temperature work in igneous fusion and ebullition chiefly in relation to pressure," *US Geol. Survey Bull.* 103 (1893), 57 pp.

2. A. L. Day, "Director's Report on the Geophysical Laboratory," *CIW Year Book* 5 (1907), 177–185, at p. 179.

3. W. Schreyer and H. S. Yoder, Jr., "Albert Ludwig, an early German pioneer in high-pressure research," *Ber. Deut. Miner. Ges.* 14 (2002), 1–4.

4. J. Johnston and L. H. Adams, "The influence of pressure on the melting points of certain metals," *Am. J. Sci.*, 4th series, 31 (1911), 501–517.

5. A. Ludwig, "On the dependence of valence upon volume in certain trivalent elements," *J. Am. Chem. Soc.* 31 (1909), 1130–1136.

6. L. H. Adams, E. D. Williamson, and J. Johnston, "The determination of the compressibility of solids at high pressures," *J. Am. Chem. Soc.* 41 (1919), 12–42.

7. F. H. Smyth and L. H. Adams, "The system calcium oxide – carbon dioxide," *J. Am. Chem. Soc.* 45 (1923), 1167–1184.

8. P. Eskola, "On pressure," *Proc. Finnish Acad. Sci. Lett.* (1958), 1–17.

9. R. W. Goranson, "The solubility of water in granite magmas," *Am. J. Sci.*, 5th series, 22 (1931), 481–502; Goranson, "Silicate-water systems: Phase equilibria in the $NaAlSi_3O_8$-H_2O and $KAlSi_3O_8$-H_2O systems at high temperatures and pressures," *Am. J. Sci.*, 5th series, 35A (1938), 71–91.

10. L. H. Adams, "The compressibility of diamond," *J. Wash. Acad. Sci.* 11 (1921), 45–50; L. H. Adams and R. E. Gibson, "The elastic properties of certain basic rocks and of their constituent minerals," *Proc. Nat. Acad. Sci.* 15 (1929), 713–724.

11. L. H. Adams and J. W. Green, "The influence of hydrostatic pressure on the critical temperature of magnetization for iron and other materials," *Phil. Mag.* 12 (1931), 361–380, at p. 380.

12. L. H. Adams, "The significance of pressure and of volume in geophysical investigations," *Cooperation in Research* (CIW Publication 501, 1938), pp. 37–47.

13. R. W. Goranson and E. A. Johnson, "The attainment of high hydrostatic pressures," *Phys. Review* 57 (1940), 845.

14. O. F. Tuttle, "A new hydrothermal quenching apparatus," *Am. J. Sci.* 246 (1948), 628–635.

15. J. Van den Heurk, "Improved hydrothermal quenching apparatus," *Bull. Geol. Soc. Am.* 64 (1953), 993–996.

16. O. F. Tuttle, "The variable inversion temperature of quartz as a possible geologic thermometer," *Am. Mineral.* 34 (1949), 723–730.

17. H. S. Yoder, Jr., "Experimental mineralogy: achievements and prospects," *Bull. Mineral.* 103 (1980), 5–26.

18. H. S. Yoder, Jr., "High-low quartz inversion up to 10,000 bars," *Trans. Am. Geophys. Union* 31 (1950), 827–835.

19. H. S. Yoder, Jr., "The jadeite problem, Parts I and II," *Am. J. Sci.* 248 (1950), 225–248; 312–334; H. S. Yoder, Jr. and C. E. Weir, "Change of free energy with pressure of the reaction nepheline + albite = 2 jadeite," *Am. J. Sci.* 249 (1951), 683–694.

20. L. Coes, Jr. "A new dense crystalline silica," *Science* 118 (1953), 131–132; Coes, "High-pressure minerals," *J. Am. Ceram. Soc.* 38 (1955), 298; Coes, "Synthesis of minerals at high pressures," in R. H. Wentorf, Jr. (ed.), *Modern Very High Pressure Techniques* (Butterworths, Washington, 1962), pp. 137–150.

21. Yoder, "Experimental mineralogy," p. 17, fig. 23.

22. F. R. Boyd and J. L. England, "Development of high-pressure apparatus," *CIW Year Book* 57 (1958), 170–173; Boyd and England, "Apparatus for phase-equilibrium measurement at pressures up to 50 kb and temperatures up to 1750°C," *J. Geophys. Res.* 65 (1960), 741–748, at p. 742, fig. 1.

23. F. Birch, E. C. Robertson, and S. P. Clark, Jr., "Apparatus for pressures of 27,000 bars and temperatures of 1400°C," *Ind. Eng. Chem.* 49 (1957), 1965–1966.

24. H. T. Hall, "Some high-pressure, high-temperature design considerations: Equipment for use at 100,000 atmospheres and 3000°C," *Rev. Sci. Instrum.* 29 (1958), 267–275; Yoder, "Experimental mineralogy."

25. H. T. Hall, "Ultra-high-pressure, high-temperature apparatus: The 'belt'." *Rev. Sci. Instrum.* 31 (1960), 125–131.

26. C. E. Weir, E. R. Lippincott, A. Van Valkenburg, and E. N. Bunting, "Infrared studies in the 1- to 15-micron region to 30,000 atmospheres." *J. Res. Nat. Bur. Stand.*, Sect. A, 63 (1959), 55–62.

27. H-k. Mao and P. M. Bell, "Design of a diamond-windowed, high-pressure cell for hydrostatic pressures in the range 1 bar to 0.5 Mbar," *CIW Year Book* 74 (1975), 402–405; Mao and Bell, "High-pressure physics: sustained static generation of 1.36 to 1.72 megabars," *Science* 200 (1978), 1145–1147.

28. In one run at 1.7 Mbar and 25 °C one of the diamonds was observed to flow plastically. On analysis it was concluded that dislocations and large amounts of nitrogen may have influenced the deformation.

29. W. T. Huntress, Jr., "The Director's Report," *CIW Year Book* 00/01 (2002), 25–38.

30. J. A. Xu, H-k. Mao, and P. M. Bell, "High-pressure ruby and diamond fluorescence: observations at 0.21 to 0.55 terapascal," *Science* 232 (1986), 1404–1406.

31. Adams, Williamson, and Johnston, "Determination of the compressibility of solids," p. 14, footnote 3.

32. The unit should not be confused with the Pascal computer programming language introduced at the same time and named after the same man Blaise Pascal (1623–1662), a French mathematician and philosopher.

33. E. Lisell, "Om tryckets inflytande på det elektriska ledningsmotståndet hos metaller samt en ny metod att mäta höga tryck," *Uppsala Universitets Årsskrift, Matematik och Naturvetenskap.* 1 (1903).

34. P. W. Bridgman, "The measurement of high hydrostatic pressure. II. A secondary mercury resistance gauge," *Proc. Am. Acad. Arts Sci.* 44 (1909), 221–251.

35. Adams, Williamson, and Johnston, "Determination of the compressibility of solids," p. 13.

36. L. H. Adams, R. W. Goranson, and R. E. Gibson, "Construction and properties of the manganin resistance pressure gauge," *Rev. Sci. Instr.* 8 (1937), 230–235.

37. R. A. Forman, G. J. Piermarini, J. D. Barnett, and S. Block, "Pressure measurement made by the utilization of ruby sharp-line luminescence," *Science* 175 (1972), 284–285.

38. H-k. Mao, P. M. Bell, J. W. Shaner, and D. J. Steinberg, "Specific volume measurements of Cu, Mo, Pd, and Ag and calibration of the ruby R1 fluorescence pressure gauge from 0.06 to 1 Mbar," *J. Appl. Phys.* 49 (1978) 3276–3283.

39. H-k. Mao and R. J. Hemley, "Experimental studies of the Earth's deep interior: accuracy and versatility of diamond cells," *Phil. Trans. Roy. Soc. London* A354 (1996), 1315–1333; C.-S. Zha, T. S. Duffy, R. T. Downs, H-k. Mao, and R. J. Hemley, "Brillouin scattering and x-ray diffraction of San Carlos olivine: direct pressure

determination to 32 GPa," *Earth Planet. Sci. Lett.* 159 (1998), 25–33; R. G. Chen, C. Liebermann, and D. J. Weidner, "Elasticity of single-crystal MgO to 8 gigapascals and 1600° kelvin," *Science* 280 (1998), 1913–1916.

40. E. C. Lloyd (ed.), "Accurate characterization of the high-pressure environment," *Nat. Bur. Standards Spec. Pub.* No. 326 (1971).

41. P. W. Bridgman, "Thermoelectromotive force, Peltier heat, and Thompson heat under pressure," *Proc. Am. Acad. Arts Sci.* 53 (1918), 267–386.

42. F. Birch, "Thermoelectric measurement of high temperatures in pressure apparatus," *Rev. Sci. Instr.* 10 (1939), 137–140.

43. P. M. Bell, J. L. England, and F. R. Boyd, "The effect of pressure on the thermal emf of the platinum/platinum 10 percent rhodium thermocouple," *Nat. Bur. Stand. Spec. Publ.* No. 326 (1971), 63–65.

4 Volatile components

1. E. T. Allen and J. K. Clement, "The role of water in tremolite and certain other minerals," *Am. J. Sci.* 26 (1908), 101–118.

2. E. Posnjak and N. L. Bowen, "The role of water in tremolite," *Am. J. Sci.* 22 (1931), 203–214; W. T. Schaller, "The chemical composition of tremolite," *Mineralogic Notes Series* 3, US Geol. Surv. Bull. 610 (1916), 133–136; B. E. Warren, "The structure of tremolite $H_2Ca_2Mg_5(SiO_3)_8$," *Zeit. Krist.* 72 (1929), 42–57.

3. G. W. Morey and C. N. Fenner, "The ternary system H_2O-K_2SiO_3-SiO_2," *J. Am. Chem. Soc.* 39 (1917), 1173–1229.

4. A. L. Day and E. S. Shepherd, "Water and volcanic activity," *Bull. Geol. Soc. Am.* 24 (1913), 573–606, at 574.

5. F. H. Smyth and L. H. Adams, "The system calcium oxide-carbon dioxide," *J. Am. Chem. Soc.* 45 (1923), 1167–1184.

6. H. E. Boeke, "Die Schmelzerscheinungen und die umkehrbare Umwandlung des calcium carbonates," *Neues Jahrb. Mineral. Geol. Abh.* 1 (1912), 91–121.

7. P. Eskola, "Ala-Satakunnan kallioperusta," *Satakunta Kotiseututkimuksia* 5 (1925), 297–334.

8. A. L. Day, "Annual Report of the Director of the Geophysical Laboratory," *CIW Year Book* 30 (1931), 75–100.

9. R. W. Goranson, "The solubility of water in granite magmas," *Am. J. Sci.* 22 (1931), 481–502; Goranson, "Silicate-water systems: phase equilibria in the $NaAlSi_3O_8$-H_2O and $KAlSi_3O_8$-H_2O systems at high temperatures and pressures," *Am. J. Sci.,* Day Volume 35-A (1938), 71–91.

10. Goranson, "Silicate-water systems."

11. G. W. Morey and E. Ingerson, "The pneumatolytic and hydrothermal alteration and synthesis of silicates," *Econ. Geol.* 32 Supplement (1937), 607–761.

12. L. H. Adams, "Systems with water under pressure," *CIW Year Book* 40 (1941), 38–41; G. W. Morey, "Hydrothermal synthesis," *J. Am. Ceram. Soc.* 36 (1953), 279–285.

13. O. F. Tuttle, "Two pressure vessels for silicate-water studies," *Bull. Geol. Soc. Am.* 60 (1949), 1727–1729; H. S. Yoder, Jr., "High-low quartz inversion up to 10,000 bars," *Trans. Am. Geophys. Union* 31 (1950), 827–835.

14. N. L. Bowen, and O. F. Tuttle, "The system MgO-SiO_2-H_2O," *Bull. Geol. Soc. Am.* 60 (1949), 439–460.

15. H. S. Yoder, Jr., "The MgO-Al$_2$O$_3$-SiO$_2$-H$_2$O system and the related metamorphic facies," *Am. J. Sci.*, Bowen Volume 250A (1952), 569–627.

16. Goranson, "Solubility of water."

17. H. S. Yoder, Jr. and C. E. Tilley, "Origin of basalt magmas: an experimental study of natural and synthetic rock systems," *J. Petrol.* 3 (1962), 342–532.

18. H. S. Yoder, Jr., "Synthetic basalt," *CIW Year Book* 53 (1954), 106–107.

19. F. R. Boyd, "Hydrothermal investigations of amphiboles." In *Researches in Geochemistry*, P. H. Abelson (ed.), (John Wiley & Sons, 1959), pp. 377–396; W. G. Ernst, "The stability relations of magnesioriebeckite," *Geochim. Cosmochim. Acta* 19 (1960), 10–40.

20. G. W. Morey and J. M. Hesselgesser, "The solubility of some minerals in superheated steam at high pressures," *Econ. Geol.* 46 (1951), 821–835.

21. H. S. Yoder, Jr., "Zeolites," *CIW Year Book* 53 (1954), 121–122.

22. H. J. Greenwood, "Water pressure and total pressure in metamorphic rocks," *CIW Year Book* 59 (1960), 58–63.

23. H. J. Greenwood, "The system NaAlSi$_2$O$_6$-H$_2$O-Argon: Total pressure and water pressure in metamorphism," *J. Geophys. Res.* 66 (1961), 3923–3946; Greenwood, "Metamorphic reactions involving two volatile components," *CIW Year Book* 61 (1962), 82–85.

24. M. Rosenhauer and D. H. Eggler, "Solution of H$_2$O and CO$_2$ in diopside melt," *CIW Year Book* 74 (1975), 474–479.

25. A. A. Kadik and D. H. Eggler, "Melt-vapor relations on the join NaAlSi$_3$O$_8$-H$_2$O-CO$_2$," *CIW Year Book* 74 (1975), 479–484.

26. D. S. Khorzhinskii, "Mobility and inertness of components in metasomatism," *Bull. Acad. Sci. USSR, Geol. Ser.* 1 (1936), 35–60; Korzhinskii, "The theory of systems with perfectly mobile components and processes of mineral formation," *Am. J. Sci.* 263 (1965), 193–205; Korzhinskii, *Theory of Metasomatic Zoning* (Oxford, Clarendon Press, 1970); J. B. Thompson, "Thermodynamic basis for the mineral facies concept," *Am. J. Sci.* 253 (1955), 65–103; A. Hofmann, "Chromatographic theory of infiltration metasomatism and its application to feldspars," *Am. J. Sci.* 272 (1972), 69–90.

27. J. J. Hemley, "Some mineralogical equilibria in the system K$_2$O-Al$_2$O$_3$-SiO$_2$-H$_2$O," *Am. J. Sci.* 257 (1959), 241–270.

28. J. D. Frantz and A. Weisbrod, "Infiltration metasomatism in the system K$_2$O-Al$_2$O$_3$-SiO$_2$-H$_2$O-HCl," *CIW Year Book* 72 (1973), 507–515; Frantz and Weisbrod, "Infiltration metasomatism in the system K$_2$O-Al$_2$O$_3$-SiO$_2$-H$_2$O-HCl." In A. W. Hofmann, B. J. Giletti, H. S. Yoder, Jr., and R. A. Yund (eds.), *Geochemical Transport and Kinetics* (Carnegie Institution of Washington Publication 634, 1974), pp. 261–271.

29. J. D. Frantz and H-k. Mao, "Bimetasomatism resulting from intergranular diffusion: multimineralic zone sequences," *CIW Year Book* 74 (1975), 417–424.

30. J. D. Frantz and H-k. Mao, "Metasomatic zoning resulting from intergranular diffusion: concentration profiles and the determination of complicated reaction paths in n-component systems," *CIW Year Book* 75 (1976), 759–771.

31. H. P. Eugster, "Reduction and oxidation in metamorphism." In P. H. Abelson (ed.), *Researches in Geochemistry* (John Wiley & Sons, 1959), pp. 397–426.

32. H. R. Shaw, "Hydrogen-water vapor mixtures: Control of hydrothermal atmospheres by hydrogen osmosis," *Science* 139 (1963), 1220–1222.

33. J. D. Frantz, J. M. Ferry, R. K. Popp, and D. A. Hewitt, "Redesign of the Shaw Apparatus for controlled hydrogen fugacity during hydrothermal experimentation," *CIW Year Book* 76 (1977), 660–662.

34. N. L. Bowen, "Diffusion in silicate melts," *J. Geol.* 29 (1921), 295–317.

35. H. S. Yoder, Jr., "Contemporaneous rhyolite and basalt," *CIW Year Book* 69 (1971), 141–145; Yoder, Jr., "Contemporaneous basaltic and rhyolitic magmas," *Am. Mineral.* 58 (1973), 153–171.

36. H. S. Yoder, Jr., "Diffusion between magmas of contrasting composition," *CIW Year Book* 70 (1971), 105–108.

5 Volcanology

1. A. Heilprin, *Mont Pelée and the Tragedy of Martinique: A Study of the Great Catastrophes of 1902, With Observations and Experiences in the Field* (J. B. Lippincott Co., Philadelphia, London, 1903).

2. H. S. Washington, *The Roman Comagmatic Region* (Carnegie Institution of Washington Publication 57, 1906), 199 pp.

3. H. S. Yoder, Jr., "Italian Volcanology: Geophysical Laboratory Contributions, 1905–1965." In N. Morello (ed.), *Volcanoes and History: Proceedings of the 20th INHIGEO Symposium* (Brigati, Genova, Italy, 1998), pp. 707–734.

4. Perret was "closely associated" with the Geophysical Laboratory and "often referred to himself as a member of the regular staff." Except for a brief period from 1904 when he served with the Italian Government Observatory, "he was never on a regular salary basis with any organization." At various times he received financial aid from friends, interested persons, and several organizations. From 1915 until 1931 he received direct assistance from the Geophysical Laboratory. During World War I he served with the American Red Cross in Italy. On 13 March 1931 he was officially designated a Research Associate of the Institution. Grants in support of his studies, administered by the Geophysical Laboratory, continued until his death in 1943. See M. Giblin, "Frank Alvord Perret (1867–1943) Necrologie," *Bull. Volcan.* 10 (1950), 191–196.

5. F. A. Perret, *The Vesuvius Eruption of 1906* (Carnegie Institution of Washington Publication 339, 1924), 151 pp.

6. H. S. Washington and A. L. Day, "Present condition of the volcanoes of Southern Italy," *Bull. Geol. Soc. Am.* 26 (1915), 375–388.

7. In the *CIW Year Book* 2 (1904, p. 183) it was suggested that a branch laboratory be established in Hawaii in order to take advantage of the opportunity to study volcanism.

8. R. B. Sosman, "Annual report of the director of the Geophysical Laboratory," *CIW Year Book* 18 (1919), 153–174.

9. A. L. Day and E. S. Shepherd, "Water and the magmatic gases," *CIW Year Book* 12 (1913), 145–146.

10. A. L. Day and E. T. Allen, *The Volcanic Activity and Hot Springs of Lassen Peak* (Carnegie Institution of Washington Publication 360, 1925), 190 pp.

11. *Ibid.*, p. 52.

12. "The Geysers" is a misnomer as no geysers occur there.

13. E. T. Allen and A. L. Day, *Steam Wells and Other Thermal Activity at 'The Geysers,' California* (Carnegie Institution of Washington Publication 378, 1927), 106 pp.

14. This was only the second time pressures in steam wells had been measured. The first measurements were attempted at the steam power plant in Larderello, Italy.

15. C. N. Fenner, "The Katmai region, Alaska, and the great eruption of 1912," *J. Geol.* 28 (1920), 569–606.

16. E. T. Allen and A. L. Day, *Hot Springs of the Yellowstone National Park* (Carnegie Institution of Washington Publication 466, 1935), 515 pp.

17. J. P. Iddings, "Geology of the Yellowstone Park," *US Geol. Surv. Mon.* 32 (1899), Pt. 2.

18. C. N. Fenner, "Contact relations between rhyolite and basalt on Gardiner River, Yellowstone Park," *Bull. Geol. Soc. Am.* 49 (1938), 1441–1484.

19. R. E. Wilcox, "Rhyolite-basalt complex on Gardiner River, Yellowstone Park, Wyoming," *Bull. Geol. Soc. Am.* 55 (1944), 1047–1080.

20. C. N. Fenner, "Rhyolite-basalt complex on Gardiner River, Yellowstone Park, Wyoming: A discussion," *Bull. Geol. Soc. Am.* 55 (1944), 1081–1096.

21. F. R. Boyd, "Welded tuffs and flows in the rhyolite plateau of Yellowstone Park, Wyoming," *Bull. Geol. Soc. Am.* 72 (1961), 387–426.

22. T. F. W. Barth, *Volcanic Geology, Hot Springs, and Geysers of Iceland* (Carnegie Institution of Washington Publication 587, 1950), 174 pp.

23. K. Muehlenbachs, "The oxygen isotope geochemistry of siliceous volcanic rocks from Iceland," *CIW Year Book* 72 (1973), 593–597.

24. E. G. Zies, "Temperature measurements at Paricutin Volcano," *Trans. Am. Geophys. Union* 27 (1946), 178–180.

25. F. Chayes and E. G. Zies, "Sanidine phenocrysts in some peralkaline volcanic rocks," *CIW Year Book* 61 (1962), 112–118.

26. E. G. Zies, "Chemical analyses of two pantellerites," *J. Petrol.* 1 (1960), 304–308; Zies, "A titaniferous basalt from the Island of Pantelleria," *J. Petrol.* 3 (1962), 177–180.

27. The "R_2O_3" group includes the principal constituents of the ammonia precipitate stage in rock analysis such as the hydroxides of aluminum and iron as well as titanium, phosphorous, vanadium, zirconium, and the rare earths in addition to partial amounts of Mn, Ni, Co and Cu. See A. W. Groves, *Silicate Analysis; A Manual for Geologists and Chemists, With Chapters on Check Calculations and Geochemical Data*, 2nd edn (George Allen & Unwin Ltd., London, 1951), 336 pp., at p. 56.

28. G. W. Morey, "Development of pressure in magma as a result of crystallization," *J. Wash. Acad. Sci.* 12 (1922), 219–230.

29. Day and Allen, *Volcanic Activity and Hot Springs*, p. 78, fig. 41b.

30. H. S. Yoder, Jr., "Diopside-anorthite-water at five and ten kilobars and its bearing on explosive volcanism," *CIW Year Book* 64 (1965), 82–89.

31. *Ibid.*

6 Thermodynamics

1. J. W. Gibbs, "On the equilibrium of heterogeneous substances," *Trans. Conn. Acad.* 3 (1876), 108–248.

2. G. W. Morey, "The phase rule and heterogeneous equilibrium." In *Commentary on the Scientific Writings of J. Willard Gibbs*, Volume 1 (Yale University Press, 1936), pp. 233–293.

3. G. Tunell, *Condensed Collections of Thermodynamic Formulas for One-component and Binary Systems of Unit and Variable Mass* (Carnegie Institution of Washington Publication 408B, 1985).

4. W. P. White, "Specific heats of silicates and platinum," *Am. J. Sci.* 28 (1909), 334–346.

5. W. P. White, *The Modern Calorimeter* (American Chemical Society Monograph 42, 1928), at p. 186.

6. D. R. Torgeson and Th. G. Sahama, "A hydrofluoric acid solution calorimeter and the determination of the heats of formation of Mg_2SiO_4, $MgSiO_3$ and $CaSiO_3$," *J. Am. Chem. Soc.* 70 (1948), 2156–2160.

7. F. C. Kracek, K. J. Neuvonen, and G. Burley, "Thermochemistry of mineral substances, I: A thermodynamic study of the stability of jadeite," *J. Wash. Acad. Sci.* 41 (1951), 373–383.

8. H. S. Roberts, "Direct measurement of silicate heats of melting," *Am. J. Sci.* 35A (1938), 273–287.

9. N. L. Bowen, "The melting phenomena of the plagioclase feldspars," *Am. J. Sci.* 35 (1913), 577–599.

10. T. W. Richards and W. N. Stull, *New Method for Determining Compressibility* (Carnegie Institution of Washington Publication 7, 1903). They suggested the word "megabar" for a pressure of a megadyne per square centimeter. The definition was intended to replace "atmosphere" and "kilogram per square centimeter." See also L. H. Adams, "A note on the compressibility with pressure," *J. Wash. Acad. Sci.* 17 (1927), 529–533.

11. G. F. Becker, "Experiments on the elasticity and plasticity of solids," *CIW Year Book* 3 (1905), 80.

12. F. D. Adams and E. G. Coker, *An Investigation into the Elastic Constants of Rocks, More Especially with Reference to Cubic Compressibility* (Carnegie Institution of Washington Publication 46, 1906).

13. G. F. Becker, "Experiments on elasticity and plasticity of solids," *CIW Year Book* 5 (1907), 175–176.

14. J. Johnston, "A correlation of the elastic behavior of metals with certain of their physical constants," *J. Am. Chem. Soc.* 34 (1912), 788–802.

15. R. E. Cohen and Z. Gong, "Melting and melt structure of MgO at high pressures," *Phys. Rev. B.* 50 (1994), 12301–12311.

16. R. E. Cohen and J. S. Weitz, "The melting curve and premelting of MgO." In M. H. Manghnani and T. Yagi (eds.), *Properties of Earth and Planetary Material at High Pressure and Temperature* (American Geophysical Union, 1998), pp. 185–96.

17. J. Ita and R. E. Cohen, "Diffusion in MgO at high pressure: Implications for lower mantle rheology," *Geophys. Res. Lett.* 25 (1998), 1095–1098.

18. F. C. Marton, J. Ita, and R. E. Cohen, "Pressure-volume-temperature equation of state of $MgSiO_3$ perovskite from molecular dynamics and constraints on lower mantle composition," *J. Geophys. Res.* 106 (2001), 8615–8627.

19. L. H. Adams, "The significance of pressure and of volume in geophysical investigations." In *Cooperation in Research* (Carnegie Institution of Washington Publication 501, 1938), pp. 37–47, at p. 47.

20. L. H. Adams, E. D. Williamson, and J. Johnston, "The determination of the compressibilities of solids at high pressures," *J. Am. Chem. Soc.* 41 (1919), 12–42.

21. L. H. Adams and E. D. Williamson, "The compressibility of minerals and rocks at high pressures," *J. Franklin Inst.* 195 (1923), 472–529.

22. L. H. Adams and R. E. Gibson, "The elastic properties of certain basic rocks and of their constituent minerals," *Proc. Nat. Acad. Sci.* 15 (1929), 713–724.

23. R. M. Hazen and L. W. Finger, "Compressibility and crystal structure of Angra dos Reis fassaite to 52 kbar," *CIW Year Book* 76 (1977), 512–515; Hazen and Finger, *Comparative Crystal Chemistry: Temperature, Pressure, Composition and Variation of Crystal Structure* (John Wiley and Sons, 1982), 231 pp.

24. B. Siemens and F. Seifert, "High-pressure wüstite: Cell parameters and Mössbauer spectra," *CIW Year Book* 78 (1979), 625–626.

25. R. M. Hazen, H-k. Mao, L. W. Finger, and P. M. Bell, "Crystal structure and compression of Ar, Ne, and CH_4 at 20 °C to 90 kbar," *CIW Year Book* 79 (1980), 348–351.

26. D. Rumble, "The adiabatic gradient and adiabatic compressibility," *CIW Year Book* 75 (1976), 651–655.

27. R. J. Hemley, H-k. Mao, L. W. Finger, A. P. Jephcoat, R. M. Hazen, and C. S. Zha, "Equation of state of solid hydrogen and deuterium from single-crystal X-ray diffraction to 26.5 GPa," *Phys. Rev. B* 42 (1990), 6458–6470.

28. P. W. Bridgman, *A Condensed Collection of Thermodynamic Formulas* (Harvard University Press, 1925).

29. R. W. Goranson, *Thermodynamic Relations in Multi-Component Systems* (Carnegie Institution of Washington Publication 408, 1930).

30. G. Tunell, *Condensed Collections of Thermodynamic Formulas for One-Component and Binary Systems of Unit and Variable Mass* (Carnegie Institution of Washington Publication 408B, 1985), 294 pp.

31. G. Tunell, *Thermodynamic Relations in Open Systems* (Carnegie Institution of Washington Publication 408A, 1977), 69 pp.

32. P. E. Gibson, "Thermodynamics and thermochemistry," *Ann. Surv. Am. Chem.* 10 (1935), 59–77.

33. L. H. Adams, "Activity and related thermodynamic quantities; their definition, and variation with temperature and pressure," *Chem. Rev.* 19 (1936), 1–26.

34. G. Tunell, "Notation for the derivatives of the two types of line integral in thermodynamics," *J. Chem. Phys.* 9 (1941), 191–192.

35. R. W. Goranson, "Heat capacity; heat of fusion," *Geol. Soc. Am. Special Paper* 36, *The Handbook of Physical Constants*, ed. F. Birch (1942), 223–242.

36. H. S. Yoder, Jr., "The jadeite problem," *Am. J. Sci.* 248 (1950), 225–248; 312–334.

37. See H. S. Yoder, Jr. and C. E. Weir, "Change of free energy with pressure of the reaction nepheline + albite = 2 jadeite," *Am. J. Sci.* 249 (1951), 683–694.

38. L. Coes, "High-pressure minerals," *J. Am. Ceram. Soc.* 38 (1955), 298.

39. E. C. Robertson, F. Birch, and G. J. F. MacDonald, "Experimental determination of jadeite stability relations to 25,000 bars," *Am. J. Sci.* 255 (1957), 115–137.

40. H. S. Yoder and C. W. Chesterman, "Jadeite of San Benito County California," *California Division of Mines and Geology Special Report* 10-C (1951), 1–8.

41. L. H. Adams, "A note on the stability of jadeite," *Am. J. Sci.* 521 (1953), 299–308.

42. Y. Fei, H-k. Mao, and B. O. Mysen, "Experimental determination of element partitioning and calculation of phase relations in the MgO-FeO-SiO_2 system at high pressure and high temperature," *J. Geophys. Res.* 96 (1991), 2157–2169.

43. T. Katsura and E. Ito, "The system Mg_2SiO_4-Fe_2SiO_4 at high pressures and temperatures: precise determination of stabilities of olivine, modified spinel and spinel," *J. Geophys. Res.* 94 (1989), 15663–15670; E. Ito and E. Takahashi, "Postspinel transformations in the system Mg_2SiO_4-Fe_2SiO_4 and some geophysical implications," *J. Geophys. Res.* 94 (1989), 10637–10646.

7 X-ray crystallography

1. G. Donnay, "Crystallography: fifty years of X-ray crystallography at the Geophysical Laboratory, 1919–1969." In D. McLachlan, Jr. and J. P. Gluskar (eds.), *Crystallography in North America* (American Crystallographic Association, 1983), pp. 37–41.

2. C. L. Burdick and J. H. Ellis, "The crystal structure of chalcopyrite determined by X-rays," *J. Am. Chem. Soc.* 39 (1917), 2518–2525.

3. R. W. G. Wyckoff, "The determination of the structure of crystals," *J. Franklin Inst.* 191 (1921), 199–230; Wyckoff, *The Analytical Expression of the Results of the Theory of Space Groups* (Carnegie Institution of Washington Publication 318, 1922), 180 pp.

4. R. W. G. Wyckoff, "The crystal structures of some carbonates of the calcite group," *Am. J. Sci.* 50 (1920), 317–320; 351–353.

5. R. W. G. Wyckoff, "The crystal structure of the high temperature form of cristobalite (SiO_2)," *Am. J. Sci.* 9 (1925), 448–459.

6. R. W. G. Wyckoff and E. Posnjak, "The crystal structure of the cuprous halides," *J. Am. Chem. Soc.* 44 (1922), 30–36.

7. From Donnay, "Crystallography: Fifty years." See F. C. Kracek, E. Posnjak, and S. B. Hendricks, "Gradual transition in sodium nitrate. II. The structure at various temperatures and its bearing on molecular rotation," *J. Am. Chem. Soc.* 53 (1931), 3339–3348; S. B. Hendricks, E. Posnjak, and F. C. Kracek, "Molecular rotation in the solid state. The variation of the crystal structure of ammonium nitrate with temperature," *J. Am. Chem. Soc.* 51 (1932), 2766–2786.

8. T. F. W. Barth and E. Posnjak, "The spinel structure: An example of variate atom equipoints," *J. Wash. Acad. Sci.* 21 (1931), 255–258.

9. G. Tunell and C. J. Ksanda, "The crystal structures of calaverite," *J. Wash. Acad. Sci.* 25 (1935), 32–33; G. Tunell and C. J. Ksanda, "The space group and unit cell of sylvanite," *Am. Miner.* 22 (1937), 728–730; J. J. Fahey and G. Tunell, "Bradleyite, a new mineral, sodium phosphate-magnesium carbonate," *Am. Miner.* 26 (1941), 646–650; G. Tunell and C. J. Ksanda, "The crystal structure of krennerite," *J. Wash. Acad. Sci.* 26 (1936), 507–509.

10. A. L. Patterson and G. Tunell, "A method for the summation of the Fourier series used in x-ray analysis of crystal structures," *Am. Miner.* 27 (1942), 655–679.

11. J. D. H. Donnay, G. Tunell, and T. F. W. Barth, "Various modes of attack in crystallographic investigation," *Am. Miner.* 19 (1934), 437–458.

12. J. Van den Heurk, "Norelco powder X-ray diffraction specimen changer," *CIW Year Book* 53 (1954), 140–141.

13. Gabrielle Donnay's maiden name was Hamburger, and she did her thesis on the structure of tourmaline under Professor M. J. Buerger. Her fellow students were greatly amused by a joint publication under their names of Hamburger and Buerger in 1948. She married Professor J. D. H. Donnay of Johns Hopkins University in

July 1949. Her productivity resulted from intense concentration on the problems at hand. She shared an office at the Geophysical Laboratory with three Fellows who were not as considerate of the need for the quiet she demanded. Exasperated, she moved her desk to the lounge adjoining the ladies lavatory!

14. G. Donnay and J. D. H. Donnay, "Crystal geometry of some alkali silicates," *Am. Miner.* 38 (1953), 163–171.

15. G. Donnay and M. J. Buerger, "The determination of the crystal structure of tourmaline," *Acta. Cryst.* 3 (1950), 5–12; G. Donnay, F. E. Seufle, A. Thorpe, and S. White, "Magnetic properties of tourmalines," *CIW Year Book* 65 (1967), 295–299.

16. G. Donnay and D. L. Pawson, "X-ray diffraction studies of echinoderm plates," *Science* 166 (1969), 1147–1150.

17. J. D. H. Donnay, W. Nowacki, and G. Donnay, "Crystal Data," *Geol. Soc. Am. Memoir* 60 (1954); J. D. H. Donnay and G. Donnay, "Crystal Data: Determinative Tables," *Am. Crystal. Assoc. Monograph* 5 (1963).

18. N. Morimoto and G. Kullerud, "Single-crystal studies of Cu_9S_5-Cu_5FeS_4 solid solutions," *CIW Year Book* 58 (1959), 201–203.

19. N. Morimoto and J. L. England, "High-temperature Buerger precession camera," *CIW Year Book* 59 (1960), 175.

20. N. Morimoto and G. Kullerud, "Polymorphism in bornite," *Am. Miner.* 46 (1961), 1270–1282; Morimoto and Kullerud, "Polymorphism in digenite," *Am. Miner.* 48 (1963), 110–123.

21. J. V. Smith and H. S. Yoder, Jr., "Experimental and theoretical studies of the mica polymorphs," *Min. Mag.* 31 (1956), 209–235.

22. C. W. Burnham, "Refinement of the crystal structure of sillimanite," *Zeit. Krist.* 118 (1963), 127–148; Burnham, "Refinement of the crystal structure of kyanite," *Zeit. Krist.* 118 (1963), 337–360.

23. C. T. Prewitt and C. W. Burnham, "The crystal structure of jadeite, $NaAlSi_2O_6$," *Am. Miner.* 59 (1966), 956–975.

24. L. Güven and C. W. Burnham, "The crystal structure of 3-T muscovite," *Zeit. Krist.* 125 (1967), 163–183.

25. L. W. Finger and C. W. Burnham, "Peak-width calculations for equi-inclination diffraction geometry," *Zeit. Krist.* 127 (1968), 101–109.

26. A. T. Anderson, Jr., T. E. Bunch, E. N. Cameron, S. E. Haggerty, F. R. Boyd, L. W. Finger, O. B. James, K. Keil, M. Prinz, P. Ramdohr, and A. El Goresy, "Armalcolite: a new mineral from the Apollo 11 samples," *Proc. Apollo 11 Lunar Sci. Conf., Geochim. Cosmochim. Acta Suppl.* 1 (1970), 55–63.

27. B. A. Wechsler, C. T. Prewitt, and J. J. Papike, "Structure and chemistry of lunar and synthetic armalcolite," *Lunar Science VI, Abstracts,* (Lunar Science Institute, Houston, 1975), pp. 860–862.

28. L. W. Finger, C. G. Hadidiacos, and Y. Ohashi, "A computer-automated, single-crystal X-ray diffractometer," *CIW Year Book* 72, 694–699; Finger and E. Prince, "A system of Fortran IV computer programs for crystal structure computations," *Natl. Bur. Stand. (US) Tech. Note* 854 (1975).

29. Finger, L. W., "Refinement of the crystal structure of an anthophyllite," *CIW Year Book* 68 (1970), 283–288.

30. R. H. McCallister, L. W. Finger, and Y. Ohashi, "Intercrystalline Fe^{2+}-Mg equilibria in three natural Ca-rich clinopyroxenes," *Am. Mineral.* 61 (1976), 671–676.

31. P. H. Nixon and F. R. Boyd, "Petrogenesis of the granular and sheared ultrabasic nodule suite in kimberlites." In P. H. Nixon (ed.), *Lesotho Kimberlites* (Lesotho National Development Corp., Maseru, 1973), pp. 48–56.

32. F. Seifert and D. Virgo, "Kinetics of the Fe^{2+}-Mg, order–disorder reaction in anthophyllites: Quantitative cooling rates," *Science* 188 (1975), 1107–1109.

33. Y. Ohashi and L. W. Finger, "Stepwise cation ordering in bustamite and disordering in wollastonite," *CIW Year Book* 75 (1976), 746–752.

34. Y. Ohashi and L. W. Finger, "An effect of temperature on the feldspar structure: crystal structure of sanidine at 800°C," *CIW Year Book* 74 (1975), 569–572; Finger and Ohashi, "The thermal expansion of diopside to 800°C and a refinement of the crystal structure at 700°C," *Am. Mineral.* 61 (1976), 303–310.

35. R. M. Hazen and L. W. Finger, "Crystal structures and compressibilities of pyrope and grossular to 60 kbar," *Am. Mineral.* 63 (1978), 297–303.

36. L. Merrill and W. Bassett, "Miniature diamond anvil pressure cell for single-crystal X-ray diffraction studies," *Rev. Sci. Instrum.* 45 (1974), 290–294.

37. L. W. Finger, R. M. Hazen, G. Zou, H-k. Mao, and P. M. Bell, "Structure and compression of crystalline argon and neon at high pressure and room temperature," *Appl. Phys.* 39 (1981), 892–894.

38. R. M. Hazen and L. W. Finger, *Comparative Crystal Chemistry* (John Wiley & Sons, New York, 1982).

39. H-k. Mao, A. P. Jephcoat, R. J. Hemley, L. W. Finger, C. S. Zha, R. M. Hazen, and D. E. Cox, "Synchrotron X-ray diffraction measurements of single-crystal hydrogen to 26.5 gigapascals," *Science* 239 (1988), 1131–1134.

40. M. A. Strzhemechny and R. J. Hemley, "New ortho-para conversion mechanism in dense solid hydrogen," *Phys. Rev. Lett.* 85 (2000), 5595–5598.

41. H-k. Mao and R. J. Hemley, "Optical studies of hydrogen above 200 gigapascals: Evidence for metallization by band overlap," *Science* 244 (1989), 1462–1465; Mao and Hemley, "Hydrogen at high pressure," *Am. Sci.* 80 (1992), 234–247.

8 Silicate liquid structure

1. B. O. Mysen, D. Virgo and C. M. Scarfe, "Relations between the anionic structure and viscosity of silicate melts – a Raman spectroscopic study," *Am. Mineral.* 65 (1980), 690–710; D. Virgo, B. O. Mysen, and I. Kushiro, "Anionic constitution of 1-atmosphere silicate melts: Implication for the structure of igneous melts," *Science* 208 (1980), 1371–1373.

2. F. Seifert, B. O. Mysen, and D. Virgo, "Structural similarity between glasses and melts relevant to petrological processes," *Geochim. Cosmochim. Acta* 45 (1981), 1879–1884.

3. D. B. Dingwell and S. L. Webb, "Relaxation in silicate melts," *Eur. J. Mineral.* 2 (1990), 427–449.

4. I. Kushiro, "On the nature of silicate melt and its significance in magma genesis: Regularities in the shift of the liquidus boundaries involving olivine, pyroxene, and silica minerals," *Am. J. Sci.* 275 (1975), 411–431.

5. W. Zachariassen, "The atomic arrangement in glass," *J. Am. Ceram. Soc.* 54 (1932), 3841–3851.

6. C. L. Babcock, "Substructures in silicate glasses," *J. Am. Ceram. Soc.* 51 (1968), 163–169.

7. B. O. Mysen, D. Virgo, and F. A. Seifert, "Structure of silicate melts: Implication for chemical and physical properties of natural magma," *Rev. Geophys.* 20 (1982), 353–383.

8. B. O. Mysen and J. D. Frantz, "Raman spectroscopy of silicate melts at magmatic temperatures: Na_2O-SiO_2, K_2O-SiO_2 and Li_2O-SiO_2 binary compositions in the temperature range 25–1475°C," *Chem. Geol.* 96 (1992), 321–332.

9. B. O. Mysen, "Structure and properties of magmatic liquids: From haplobasalt to haploandesite," *Geochim. Cosmochim. Acta* 63 (1999), 95–112; Mysen, "Physics and chemistry of silicate glasses and melts," *Eur. J. Mineral.* 15 (2003), 781–802.

10. F. Seifert, B. O. Mysen, and D. Virgo, "Three-dimensional network structure of quenched melts (glass) in the systems $SiO_2-NaAlO_2$, $SiO_2-CaAl_2O_4$ and $SiO_2-MgAl_2O_4$," *Am. Mineral.* 67 (1982), 696–717.

11. B. O. Mysen, "Magmatic silicate melts: Relations between bulk composition, structure and properties," *Geochem. Soc. Special Pub.* 1 (1987), 375–399.

12. B. O. Mysen, F. J. Ryerson, and D. Virgo, "The structural role of phosphorus in silicate melts," *Am. Mineral.* 66 (1981), 106–117.

13. B. O. Mysen and D. Neuville, "Effect of temperature and TiO_2 content on the structure of $Na_2Si_2O_5-Na_2Ti_2O_5$ melts and glasses," *Geochim. Cosmochim. Acta* 59 (1995), 325–342; H. V. Alberto, B. O. Mysen, and N. DeCampos, "The structural role of titanium in silicate glasses: A Raman study of the system $CaO-SiO_2-TiO_2$," *Phys. Chem. Glasses* 36 (1995), 114–122.

14. Kushiro, "On the nature of silicate melt."

15. E. B. Watson, "Two-liquid partition coefficients experimental data and geochemical implications," *Contrib. Min. Petrol.* 56 (1976), 119–134.

16. F. S. Ryerson and P. C. Hess, "The role of P_2O_5 in silicate melts," *Geochim. Cosmochim. Acta* 44 (1980), 611–625.

17. G. D. Cody, B. O. Mysen, G. Sághi-Szabó, and J. A. Tossell, "Silicate-phosphate interactions in silicate glasses and melts: I. A multinuclear (^{27}Al, ^{29}Si, ^{31}P) MAS-NMR and ab initio chemical shielding (^{31}P) study of phosphorous speciation in silicate glasses," *Geochim. Cosmochim. Acta* 65 (2001), 2395–2411; B. O. Mysen and G. D. Cody, "Silicate-phosphate interactions in silicate glasses and melts: II. Quantitative, high-temperature structure of P-bearing alkali aluminosilicate melts," *Geochim. Cosmochim. Acta* 65 (2001), 2413–2431.

18. B. O. Mysen and D. Virgo, "Influence of pressure, temperature, and bulk composition on melt structures in the system $NaAlSi_2O_2-NaFe^{3+}Si_2O_6$, *Am. J. Sci.* 278 (1978) 1307–1322.

19. B. O. Mysen, F. Seifert, and D. Virgo, "Structure and redox equilibria of iron-bearing silicate melts," *Am. Mineral.* 65 (1980), 867–884.

20. B. O. Mysen and D. Virgo, "Redox equilibria, structure, and properties of Fe-bearing aluminosilicate melts: relationships between temperature, composition, and oxygen fugacity in the system $Na_2O-Al_2O_3-SiO_2-Fe-O$," *Am. Mineral.* 74 (1989), 58–76.

21. B. O. Mysen and D. Virgo, "Trace element partitioning and melt structure: an experimental study at 1 atm. pressure," *Geochim. Cosmochim. Acta* 44 (1980), 1917–1930.

22. Mysen, Seifert, and Virgo, "Redox equilibria."

23. B. O. Mysen and J. D. Frantz, "Structure of silicate melts at high temperature: in-situ experiments in the system $BaO-SiO_2$ to 1669 °C," *Am. Mineral.* 78 (1993), 699–709.

24. J. D. Frantz and B. O. Mysen, "Raman spectra and structure of BaO-SiO_2, SrO-SiO_2, and CaO-SiO_2 melts to $1600°C$," *Chem. Geol.* 121 (1995), 155–176.

25. B. O. Mysen and D. Virgo, "Volatiles in silicate melts at high pressure and temperature: I. Interaction between OH groups and Si^{4+}, Al^{3+}, Na^+, and H^+," *Chem. Geol.* 57 (1986), 303–331.

26. B. O. Mysen and D. Virgo, "Solubility mechanisms of carbon dioxide in silicate melts: a Raman spectroscopic study," *Am. Mineral.* 65 (1980), 885–899.

27. B. O. Mysen and D. Virgo, "Structure and properties of fluorine-bearing aluminosilicate melts: the system Na_2O-Al_2O_3-SiO_2-F at 1 atm," *Contrib. Mineral. Petrol.* 91 (1985), 77–85.

28. See, for example, B. O. Mysen and R. F. Popp, "Solubility of sulfur in $CaMgSi_2O_6$ and $NaAlSi_3O_8$ melts at high pressure and temperature with controlled f_{O_2} and f_{S_2}," *Am. J. Sci.* 280 (1980), 78–92.

29. B. O. Mysen, "Solubility of volatiles in silicate melts under the pressure and temperature conditions of partial melting in the upper mantle," *Oregon Dept. Geol. Miner. Ind. Bull.* 96 (1977), 1–14.

30. B. O. Mysen and M. G. Seitz, "Trace element partitioning determined by beta-track mapping – an experimental study using carbon and samarium as examples," *J. Geophys. Res.* 80 (1975), 2627–2635; Mysen and Arculus, unpubl. data.

9 Ore geochemistry

1. The word "ore" has been used generally for any material that can be extracted by mining and has economic value. For some, the word implies extraction of only metals, whereas others include any industrial mineral, metallic (e.g. chalcopyrite, galena, sphalerite) or nonmetallic (e.g. halite, bauxite, gem stones, asbestos). Still others use the word for opaque minerals from which metals are extracted. The word can also be extended to include whole rocks, such as granite, limestone, and salt, which can be used commercially in toto. Although the word was used specifically in the early days at the Geophysical Laboratory to mean sulfides, arsenides, and fluorides, a broader view evolved to include the entire array of economic materials derived from mining through extractive processing. C. R. Van Hise, "Report on Geophysics," *CIW Year Book* 2 (1903), 173–184.

2. E. T. Allen and J. Johnston, "The exact determination of sulfur in soluble sulfates," *J. Am. Chem. Soc.* 32 (1910), 588–617; Allen and Johnston, "The exact determination of sulfur in pyrite and marcasite," *J. Indust. Engin. Chem.* 2 (1910), 1–19.

3. E. T. Allen, "Geochemistry: studies in ore deposition with special reference to the sulfides of iron," *J. Wash. Acad. Sci.* 1 (1911), 170–177; Allen, J. L. Crenshaw, J. Johnston, and E. S. Larsen, "The mineral sulfides of iron," *Am. J. Sci.* 33 (1912), 169–236.

4. E. T. Allen, J. L. Crenshaw and H. E. Merwin, "The sulfides of zinc, cadmium and mercury; their crystalline forms and genetic conditions," *Am. J. Sci.* 34 (1921), 341–396.

5. E. Posnjak, E. T. Allen, and H. E. Merwin, "The sulfides of copper," *Econ. Geol.* 10 (1915), 491–535; E. G. Zies, E. T. Allen, and H. E. Merwin, "Some reactions involved in secondary copper sulfide enrichment," *Econ. Geol.* 11 (1916), 407–503.

6. E. T. Allen and R. H. Lombard, "A method for the determination of dissociation pressures of sulfides, and its application to covellite (CuS) and pyrite (FeS_2)," *Am. J. Sci.* 43 (1917), 175–195.

7. E. T. Allen and E. G. Zies, "A chemical study of the fumeroles of the Katmai Region," *Nat. Geogr. Soc. Contributed Technical Papers, Katmai Series* No. 2 (1923), 75–155.

8. C. N. Fenner, "Pneumatolytic processes in the formation of minerals and ores," In *Ore Deposits of the Western States* (American Institute of Mining and Metallurgical Engineers, 1933), pp. 58–106.

9. N. L. Bowen, "The broader story of magmatic differentiation, briefly told" In *Ore Deposits of the Western States* (American Institute of Mining and Metallurgical Engineers, 1933), pp. 106–28.

10. H. E. Merwin and R. H. Lombard, "The system, Cu-Fe-S," *Econ. Geol.* 32 (1937), 203–284.

11. R. Brett, "Experimental data from the system Cu-Fe-S and their bearing on exsolution textures in ores," *Econ. Geol.* 59 (1964), 1241–1269.

12. The Carnegie Institution of Washington provided William Campbell, Columbia University, with several grants for apparatus to develop these techniques, particularly for applications to the metallography of steel and iron; W. Campbell, "The microscopic examination of opaque minerals," *Econ. Geol.* 1 (1906), 751–766.

13. Merwin and Lombard, "The system, Cu-Fe-S."

14. G. Kullerud, "Sulfide studies." In P. H. Abelson (ed.), *Researches in Geochemistry 2* (John Wiley & Sons, New York, 1967), pp. 286–321.

15. G. Kullerud, "The FeS-ZnS system: a geological thermometer," *Norsk. Geol. Tidskr.* 32 (1953), 61–147.

16. G. Kullerud, "Experimental techniques in dry sulfide research." In G. C. Ulmer (ed.), *Research Techniques for High Pressure and High Temperature* (Springer-Verlag, 1971), pp. 289–315.

17. G. Kullerud and H. S. Yoder, Jr., "Pyrite stability relations in the Fe-S system," *Econ. Geol.* 54 (1959), 533–572.

18. G. Kullerud, "Thermal stability of pentlandite," *Am. Mineral.* 7 (1963), 353–366.

19. J. R. Craig and G. Kullerud, "Phase relations in the Cu-Fe-Ni-S system and their application to magmatic ore deposits," *Magmatic Ore Deposits: A Symposium* (Economic Geology Monograph 4, 1969), pp. 344–358.

20. G. W. Morey and J. M. Hesselgesser, "The solubility of some minerals in superheated steam at high pressures," *Econ. Geol.* 46 (1951), 821–835.

21. H. L. Barnes, "The system ZnS-H_2S-H_2O," *CIW Year Book* 58 (1959), 163–167.

22. H. L. Barnes and G. Kullerud, "Equilibria in sulfur-containing aqueous solutions, in the system Fe-S-O, and their correlation during ore deposition," *Econ. Geol.* 56 (1961), 648–688.

23. A. J. Naldrett, "A portion of the system Fe-S-O between 900 and 1080°C and its application to sulfide ore magmas," *J. Petrol.* 10 (1969), 171–201.

24. G. W. Morey and E. Ingerson, "The pneumatolytic and hydrothermal alteration and synthesis of silicates," *Econ. Geol.* 52 (1937), 607–761.

25. G. Kullerud and H. S. Yoder, Jr., "Sulfide-silicate relations," *CIW Year Book* 64 (1965), 192–193; W. Schreyer and G. Kullerud, "Conditions of metamorphism of the Badenmais Sulfide Deposit and its country rock," *CIW Year Book* 59 (1960), 133–135.

26. G. Kullerud, "Sulfide-carbonate reactions," *CIW Year Book* 64 (1965), 188–192, at 178.

27. *Ibid.*

28. T. N. Irvine, D. W. Keith and S. G. Todd, "The J-M platinum–palladium reef of the Stillwater Complex, Montana: II Origin by double-diffusive convective magma mixing and implications for the Bushveld Complex," *Econ. Geol.* 78 (1983), 1287–1334.

29. T. N. Irvine, J. C. O. Andersen, and C. K. Brooks, "Included blocks (and blocks within blocks) in the Skaergaard intrusion: geologic relations and the origins of rhythmic modally graded layers," *Geol. Soc. Am. Bull.* 110 (1998), 1398–1447.

30. N. Z. Boctor and R. H. McCallister, "Kinetics and mechanism of the metacinnabar to cinnabar transition," *CIW Year Book* 78 (1979), 582–585.

31. J. Li, Y. Fei, H-k. Mao, K. Hirose, and S. R. Shieh, "Sulfur in the Earth's inner core," *Earth Planet. Sci. Lett.* 193 (2001), 509–514.

32. P. B. Barton, "Minerals: Building blocks of civilization," *US Geol. Surv. Rep.* (2000).

33. H. S. Yoder, Jr., "Strategic minerals: A critical research need and opportunity," *Proc. Am. Phil. Soc.* 126 (1982), 229–241; Yoder, Jr., "Geology: Significant component of new multidisciplinary sciences," *Proc. Am. Phil. Soc.* 146 (2002), 37–55.

10 Field studies

1. L. H. Adams, "Annual Report of the Acting Director of the Geophysical Laboratory," *CIW Year Book* 36 (1937), 109–134, at 109.

2. F. A. Perret, *The Volcano-Seismic Crisis at Montserrat* (Carnegie Institution of Washington Publication 512, 1939).

3. R. F. Griggs, "The great hot mud flow of the Valley of Ten Thousand Smokes," *Ohio J. Sci.* 19 (1918), 117–142.

4. C. N. Fenner, "The origin and mode of emplacement of the great tuff deposit in the Valley of Ten Thousand Smokes," *Nat. Geogr. Soc. Contrib. Tech. Papers, No. 1 Katmai Series* (1923); Fenner, "The Katmai magmatic province," *J. Geol.* 34 (1926), 673–772.

5. Fenner, "The Katmai magnetic province," p. 741.

6. H. S. Yoder, Jr., "Contemporaneous basaltic and rhyolitic magmas," *Am. Mineral.* 58 (1973), 153–171.

7. E. G. Zies, "The Valley of Ten Thousand Smokes: 1, The fumarolic incrustations and their bearing on ore deposition. 2, The acid gases contributed to the sea during volcanic activity," *Nat. Geog. Soc. Contrib. Tech. Papers, Vol. 1, No. 4 Katmai Series* (1929).

8. E. G. Zies, "Exploratory and comparative studies," *CIW Year Book* 39 (1940), 44–45; Zies, "Temperature measurements at Parícutin Volcano," *Trans. Am. Geophys. Union* 26 (1946), 178–180.

9. E. T. Allen, "Chemical aspects of volcanism with a collection of the analyses of volcanic gases," *J. Franklin Inst.* 193 (1922), 29–80.

10. H. S. Washington, "Petrology of the Hawaiian Islands: I. Kohala and Mauna Kea, Hawaii," *Am. J. Sci.* 5 (1923), 465–502; Washington, "Petrology of the Hawaiian Islands: II. Hualalai and Mauna Loa," *Am. J. Sci.* 6 (1923), 100–126; Washington, "Petrology of the Hawaiian Islands: III. Kilauea and general petrology of Hawaii," *Am. J. Sci.* 6 (1923), 338–367; Washington, "Petrology of the Hawaiian Islands: IV. The formation of aa and pahoehoe," *Am. J. Sci.* 6 (1923), 409–423; Washington and M. G. Keyes, "Petrology of the Hawaiian Islands: V. The Leeward Islands," *Am. J. Sci.* 12 (1926), 336–352.

11. T. F. W. Barth, "Mineralogical petrography of Pacific lavas," *Am. J. Sci.* 21 (1931), 377–530.
12. H. S. Yoder, Jr. and C. E. Tilley, "Origin of basaltic magma: an experimental study of natural and synthetic rock systems," *J. Petrol.* 3 (1962), 342–532.
13. L. R. Wager and W. A. Deer, "Geological investigations in East Greenland. Part III, The petrology of the Skaergaard intrusion, Kangerdlagssuaq, East Greenland," *Medd. Grønland* 105 (1939), 1–346.
14. A. R. McBirney, "Differentiation of the Skaergaard Intrusion," *Nature* 253 (1975), 691–694.
15. T. N. Irvine, "Petrology of the Duke Island Ultramafic Complex, Southeastern Alaska," *Geol. Soc. Am. Memoir* 138 (1974), 244 pp.
16. T. N. Irvine and C. H. Smith, "The ultramafic rocks of the Muskox intrusion." In P. J. Wyllie (ed.), *Ultramafic and Related Rocks* (John Wiley and Sons, New York, 1967), pp. 38–49.
17. S. G. Todd, D. J. Schissel, and T. N. Irvine, "Lithostratigraphic variations associated with the platinum-rich zone of the Stillwater Complex," *CIW Year Book* 78 (1979), 461–468.
18. For the unique and innovative large-scale experiments undertaken by Irvine in the laboratory demonstrating the products of density currents tracing the paths of the particles, see T. N. Irvine, "Density current models of tectonic processes," *CIW Year Book* 78 (1979), 450–461.
19. T. N. Irvine, J. C. Ø. Anderson, and C. K. Brooks, "Included blocks (and blocks within blocks) in the Skaergaard intrusion: geologic relations and the origins of rhythmic modally graded layers," *Geol. Soc. Am. Bull.* 110 (1998), 1398–1447.
20. J. D. Hoover, "Petrology of the marginal border series of the Skaergaard intrusion," *J. Petrol.* 30 (1989), 399–439.
21. H. R. Naslund, "Petrology of the Upper Border Series of the Skaergaard intrusion, East Greenland," *J. Petrol.* 25 (1984), 185–212.
22. L. R. Wager and G. M. Brown, *Layered Igneous Rocks* (Oliver and Boyd, Edinburgh, 1967).
23. H. C. Lewis, "On a diamondiferous peridotite, and the genesis of diamond," *Geol. Mag.* 4 (1887), 22–24.
24. B. T. C. Davis and F. R. Boyd, "The join $Mg_2Si_2O_6$–$CaMgSi_2O_6$ at 30 kb pressure and its application to pyroxenes from kimberlite," *J. Geophys. Res.* 71 (1966), 3567–3576.
25. I. D. MacGregor, "The system MgO-Al_2O_3-SiO_2: Solubility of Al_2O_3 in enstatite for spinel and garnet peridotite compositions," *Am. Mineral.* 59 (1974), 110–119.
26. G. L. Davis, "Zircons from the mantle," *CIW Year Book* 77 (1978), 895–897.
27. F. R. Boyd and J. L. England, "Pyrope," *CIW Year Book* 58 (1959), 83–87.
28. F. R. Boyd and P. H. Nixon, "Origins of the ultramafic nodules from some kimberlites of Northern Lesotho and the Monastery Mine, South Africa," *Phys. Chem. Earth* 9 (1975), 431–454.
29. F. R. Boyd, J. J. Gurney, and S. H. Richardson, "Evidence for a 150–200 km thick Archean lithosphere from diamond inclusion thermobarometry," *Nature* 315 (1985), 387–389.
30. C. S. Piggot, "Core samples of the ocean bottom and their significance," *Sci. Monthly* 46 (1938), 201–217.

31. H. Pettersson, "Teneur en radium des dépots de mer profonde," *Fascicule LXXXI des Résultats de Campagnes Scientifiques Albert Ier Prince Souvrain de Monaco*, 1930.

32. C. S. Piggot and W. D. Urry, "Time relations in ocean sediments," *Bull. Geol. Soc. Am.* 53 (1942), 1167–1210.

33. C. S. Piggot, "Radium content of ocean-bottom sediments," *Am. J. Sci.* 25 (1933), 229–238.

34. F. D. Adams, W. Cross, J. P. Iddings, L. F. Kemp, A. C. Lane, L. V. Pirsson, H. S. Washington, and John E. Wolff, "Geophysical investigations suggested," *CIW Year Book* 2 (1906), 195–201.

35. R. S. Woodward, C. Barus, T. C. Chamberlin, A. A. Michelson, C. R. Van Hise, and C. D. Walcott, "Report of Advisory Committee on Geophysics," *CIW Year Book* 1 (1903), 26–43.

36. J. Johnston and P. Niggli, "The general principles underlying metamorphic processes," *J. Geol.* 21 (1913), 481–516.

37. H. H. Hess, "An appreciation." In A. E. J. Engel, H. L. James, and B. F. Leonard (eds.), *Petrologic Studies: A Volume in Honor of A. F. Buddington* (Geological Society of America, 1962) pp. vii–xi.

38. P. Eskola, "On contact phenomena between gneiss and limestone in western Massachusetts," *J. Geol.* 30 (1922), 265–294.

39. N. L. Bowen, "Progressive metamorphism of siliceous limestone and dolomite," *J. Geol.* 48 (1940), 225–274. He composed a poem in steps of increasing temperature to aid his students: *T*remble *fo*r *d*ire *p*eril *w*alks, *m*onstrous *a*crimonies *s*purn *m*ercy's *l*aws. A further addition was later made with rankinite at the highest temperature; see F. J. Turner, "Mineralogical and structural evolution of the metamorphic rocks," *Geol. Soc. Am. Memoir* 30 (1948), 342 pp.

40. T. F. W. Barth, "Structural and petrologic studies in Dutchess County, New York, Part II: Petrology and metamorphism of the Paleozoic rocks," *Bull. Geol. Soc. Am.* 47 (1936), 775–850.

41. H. S. Yoder, Jr., "The $MgO-Al_2O_3-SiO_2-H_2O$ system and the related metamorphic facies," *Am. J. Sci.*, Bowen Volume 250A (1952), 569–627; Yoder, Jr., "Role of water in metamorphism," *Geol. Soc. Am. Special Paper* 62 (1955), 505–524.

42. C. E. Tilley, "Metamorphic zones in the southern Highlands of Scotland," *Geol. Soc. London Quart. J.* 81 (1925), 100–112; G. Barrow, "On an intrusion of muscovite-biotite gneiss in the south-east Highlands of Scotland," *Geol. Soc. London Quart. J.* 69 (1893), 330–358; C. E. Tilley, "The facies classification of metamorphic rocks," *Geol. Mag.* 61 (1924), 167–171.

43. A. R. Alderman, "Eclogites from the vicinity of Glenelg, Inverness-shire," *Geol. Soc. London Quart. J.* 92 (1936), 488–530.

44. T. C. Phemister, "Dalradian metamorphism and structure, Stonehaven to Aberdeen," *Rep. 21st Intern. Geol. Congr.* Part 13 (1960), 352–361.

45. V. M Goldschmidt, "Die Injektionsmetamorphose im Stavanger-Gebrete," *Kristiania Vidensk. Skr., Mat.-Naturv. Kl.* 10 (1921).

46. P. Eskola, "On the petrology of the Orijarvi region in south-western Finland," *Bull. Comm. Géol. Finlande* 40 (1914), 225–233; H. V. Tuominen and T. Mikkola, "Metamorphic Mg-Fe enrichment in the Orijarvi region as related to folding," *Bull. Comm. Géol. Finlande* 150 (1950), 67–92.

47. Yoder, "The $MgO-Al_2O_3-SiO_2-H_2O$ system."

48. Tuominen and Mikola, "Metamorphic Mg-Fe enrichment."

49. T. F. W. Barth, *Theoretical Petrology*, 2nd edn (New York, John Wiley and Sons, 1962), pp. 333–337.

50. H. S. Yoder, Jr., "Isograd problems in metamorphosed iron-rich sediments," *CIW Year Book* 56 (1957), 232–237.

51. H. L. James, "Sedimentary facies of iron-formation," *Econ. Geol.* 49 (1951), 253–293.

52. Bowen, "Progressive metamorphism."

53. J. B. Thompson, Jr., P. Robinson, N. G. Clifford, and N. J. Trask, "Nappes and gneiss domes in west-central New England." In E-A. Zen *et al.* (eds.), *Studies of Appalachian Geology* (John Wiley and Sons, New York, 1968), pp. 203–218.

54. D. Rumble, "Chloritoid staurolite quartzites from the moosilauke quadrangle, New Hampshire," *CIW Year Book* 69 (1971), 290–294.

55. D. Rumble, "Fe-Ti oxide minerals and the behavior of oxygen during regional metamorphism," *CIW Year Book* 70 (1971), 157–165.

56. D. Rumble, "Metasomatism in relatively closed and open systems," *CIW Year Book* 73 (1974), 371–380.

57. D. Rumble, K. Muehlenbachs, and T. C. Hoering, "Oxygen isotope geochemistry of the clough formation, Black Mountain, New Hampshire," *CIW Year Book* 74 (1975), 436–438.

58. D. Rumble and T. C. Hoering, "Rock permeability during metamorphism," *CIW Year Book* 78 (1979), 608–611.

59. D. Rumble, J. M. Ferry, T. C. Hoering, and A. J. Boucot, "Fluid flow during metamorphism at the Beaver Brook fossil locality, New Hampshire," *Am. J. Sci.* 282 (1982), 869–919.

60. A. J. Boucot and D. Rumble, "Devonian brachiopods from the sillimanite zone, Mount Moosilauke, New Hampshire," *Science* 201 (1978), 348–349.

61. D. Rumble and T. C. Hoering, "Carbon isotope geochemistry of graphite vein deposits from New Hampshire, USA," *Geochim. Cosmochim. Acta* 50 (1986), 1239–1247; Rumble, E. F. Duke, and T. C. Hoering, "Hydrothermal graphite in New Hampshire: Evidence of carbon mobility during regional metamorphism," *Geology* 14 (1986), 452–455.

62. C. P. Chamberlain and D. Rumble, "Thermal anomalies in a regional metamorphic terrane: an isotopic study of the role of fluids," *J. Petrol.* 29 (1988), 1215–1232.

63. T. F. Yui, D. Rumble, and C. H. Lo, "Unusually low delta ^{18}O ultrahigh-pressure metamorphic rocks from the Sulu terrain, eastern China," *Geochim. Cosmochim. Acta* 59 (1995), 2859–2864.

64. D. Rumble and T. F. Yui, "The Qinglongshan oxygen and hydrogen isotope anomaly near Donghai in Jiangsu province, China," *Geochim. Cosmochim. Acta* 62 (1998), 3307–3321.

65. D. Rumble, D. Giorgis, T. Ireland, Z. Zhang, H. Xu, T. F. Yui, J. Yang, Z. Xu, and J. G. Liou, "Low delta ^{18}O zircons, U-Pb dating, and the age of the Qinglongshan oxygen and hydrogen isotope anomaly near Donghai in Jiangsu Province, China," *Geochim. Cosmochim. Acta* 66 (2002), 2299–2306.

66. M. L. F. Estep and S. A. Macko, "Nitrogen isotope biochemistry of thermal springs," *Org. Geochem.* 6 (1984), 779–785.

67. L. A. Cifuentes, J. H. Sharp, and M. L. Fogel, "Stable carbon and nitrogen isotope biogeochemistry in the Delaware Estuary," *Limnol. Oceanogr.* 33 (1988), 1102–1115.

68. L. A. Cifuentes, M. L. Fogel, J. R. Pennock, and J. H. Sharp, "Biogeochemical factors that influence the stable nitrogen isotope ratio of dissolved ammonium in the Delaware Estuary," *Geochim. Cosmochim. Acta* 53 (1989), 2713–2721.

69. D. J. Velinsky, M. L. Fogel, J. F. Todd and B. M. Tebo, "Isotopic fractionation of dissolved ammonium at the oxygen–hydrogen sulfide interface in anoxic waters," *Geophys. Res. Lett.* 18 (1991), 649–652.

70. G. E. Bebout and M. L. Fogel, "Nitrogen-isotope compositions of metasedimentary rocks in the Catalina Schist, California: Implications for metamorphic devolatilization history," *Geochim. Cosmochim. Acta* 56 (1992), 2839–2849.

71. M. L. Fogel, C. Aguilar, R. Cuhel, D. J. Hollander, J. D. Willey, and H. W. Paerl, "Biological and isotopic changes in coastal waters induced by Hurricane Gordon," *Limnol. Oceanogr.* 44 (1999), 1359–1369.

72. C. Aguilar, M. L. Fogel, and H. W. Paerl, "Applications of stable nitrogen isotopes for studying atmosphere deposition impact in coastal and open ocean environments." In J. O. Grimalt and C. Dorronsoro (eds.), *Organic Geochemistry: Developments and Applications to Energy, Climate, Environment and Human History* (Iberian Association of Environmental Geochemistry, Donostia-San Sebastian, Spain, 1995), pp. 662–664.

73. H. W. Paerl, C. Aguilar, and M. L. Fogel, "Atmospheric nitrogen deposition in estuarine and coastal waters: Biogeochemical and water quality input." In J. E. Baker (ed.), *Atmospheric Deposition of Contaminants to Great Lakes and Coastal Waters* (SETAC Press, Pensacola, FL, 1997), pp. 415–429.

74. M. L. Fogel and G. Miller, "The Director's introduction," *CIW Year Book* 98 (1999), 90–91.

75. P. E. Hare, "Field studies," *CIW Year Book* 74 (1975), 629.

76. Dr. P. Edgar Hare was one of two Adventist geologists who were selected for advanced training because the Seventh-day Adventist Church lacked a person with a Ph.D. in geology or paleontology. When his research on marine shells yielded much greater ages than traditionally accepted by the Adventists, he candidly advised them of the conflict with the interpretation of the Bible. Hare was notified that he was free to remain at the Geophysical Laboratory, but would no longer be supported by the Church. The Adventist emphasis on literal Creation in six days and on the fossil-forming Noah flood emerged as the foundation of "creation science." Hare remained at the Laboratory for a further twenty-three years until retirement and produced outstanding research, including additional age determinations on ancient human skeletons. See R. L. Numbers, *Darwinism Comes to America* (Harvard University Press, Cambridge, 1998).

11 Biogeochemistry

1. The term "biogeochemistry" was first used by V. I. Vernadsky in 1924 to describe the interrelationships of biology, geology, and chemistry. The Biogeochemical Laboratory of the Academy of Sciences of the USSR was founded in 1929. The term "bioinorganic chemistry" was first used by Tsuneto Azuma in a book in Japanese entitled *Seibutsu muki kagaku*, kindly translated by Professor Ikuo Kushiro, University of Tokyo. Philip Abelson preferred the term "paleobiochemistry," whereas others preferred "organic geochemistry." As the origin-of-life problem came into focus, the terms "exobiology," "bioastronomy" (established as Commission 51 of

the International Astronomical Union in 1982), and "astrobiology" became widely used when the NASA program was instituted in 1996. See V. I. Nernadsky, *La Géochimie* (Librairie Félix Alcan, Paris, 1924), 403pp., at p. 7; T. Azuma, *Seibutsu muki kagaku* (Kanehara, Tokyo, 1929), 337pp.; P. H. Abelson, "Paleobiochemistry," *Scientific American* 195 (1956), 83–92; J. Lederberg, "Exobiology: Approaches to life beyond the Earth," *Science* 132 (1960), 393–400.

2. Dr. Philip H. Abelson was an experimental nuclear physicist on the staff of the Department of Terrestrial Magnetism (DTM), and was part of the team on biophysics. In 1955 he contributed to the *Studies of Biosynthesis in Escherichia coli*. Abelson became president of the Carnegie Institution of Washington in 1971 and was aware of the enormous growth of biophysics. When the DTM no longer was able to sustain its unique pioneering role because of retirements, Abelson took steps to dissolve the group at DTM. R. B. Roberts, P. H. Abelson, D. B. Cowie, E. T. Bolton, and R. J. Britten, *Studies of Biosynthesis in Escherichia coli* (Carnegie Institution of Washington Publication 607, 1955), 521pp.

3. W. P. Woodring, "Conference on biochemistry, paleoecology, and evolution," *Proc. Nat. Acad. Sci.* 40 (1954), 219–224.

4. P. H. Abelson, "Organic geochemistry: kerogen," *CIW Year Book* 58 (1959), 181–185.

5. P. H. Abelson, "Organic constituents of fossils," *Geol. Soc. Am. Memoir* 67 (1957), 87–92.

6. P. E. Hare, T. C. Hoering, and K. King, Jr., *Biogeochemistry of Amino Acids* (John Wiley and Sons, New York, 1980), 558pp.

7. P. E. Hare, "Amino acid dating – limitations and potential," *Geol. Soc. Am., Abstr. Prog.* 9 (1977), 1004–1005.

8. P. E. Hare and T. C. Hoering, "Separation of amino acid optical isomers by gas chromatography," *CIW Year Book* 72 (1973), 690–694.

9. Hare, "Amino acid dating."

10. P. E. Hare and G. H. Miller, "Organic geochemical studies in Spitsbergen with a field-portable, high-performance liquid chromatograph," *CIW Year Book* 79 (1980), 403–406.

11. Abelson, "Paleobiochemistry."

12. M. H. Engel, J. E. Lumberge, and B. Nagy, "Kinetics of amino acid racemization in *Sequoiadendron giganteum* heartwood," *Anal. Biochem.* 82 (1977), 415–422.

13. J. F. Wehmiller and D. F. Belknap, "Alternative kinetic models for the interpretation of amino acid eoantiomeric ratios in Pleistocene mollusks: examples from California, Washington, and Florida," *Quat. Res.* 9 (1978), 330–348.

14. J. E. Zumberge, M. H. Engel, and B. Nagy, "Amino acids in Bristlecone Pine: An evaluation of factors affecting racemization rates and paleothermometry." In P. E. Hare, T. C. Hoering, and K. King, Jr. (eds.), *Biogeochemistry of Amino Acids* (John Wiley and Sons, New York, 1980), pp. 503–525.

15. O. Nier and E. A. Gulbransen, "Variations in the relative abundance of the carbon isotopes," *J. Am. Chem. Soc.* 61 (1939), 697–698.

16. P. H. Abelson and T. C. Hoering, "The biogeochemistry of the stable isotopes of carbon," *CIW Year Book* 59 (1960), 158–165.

17. M. L. F. Estep, "Carbon and hydrogen isotopic compositions of algae and bacteria from hydrothermal environments, Yellowstone National Park," *Geochim. Cosmochim. Acta* 48 (1984), 591–599.

18. M. Dole, "The relative atomic weight of oxygen in water and in air," *J. Am. Chem. Soc.* 57 (1935), 2731.

19. R. D. Guy, M. L. Fogel, J. A. Berry and T. C. Hoering, "Isotopic fractionation during oxygen production and consumption by plants." In J. Biggins (ed.), *Progress in Photosynthesis Research* (Martinus Nijhoff Publishers, Dordrecht, 1987), pp. 597–600.

20. S. A. Macko, M. L. Fogel, P. E. Hare, and T. C. Hoering, "Isotopic fractionation of nitrogen and carbon in the synthesis of amino acids by microorganisms," *Chem. Geol.* 65 (1987), 79–92.

21. P. E. Hare, M. L. Fogel, T. W. Stafford, Jr., A. D. Mitchell and T. C. Hoering, "The isotopic composition of carbon and nitrogen in individual amino acids isolated from modern and fossil proteins," *J. Arch. Sci.* 18 (1991), 277–292.

22. G. E. Bebout and M. L. Fogel, "Nitrogen-isotope compositions of metasedimentary rocks in the Catalina Schist, California: Implications for metamorphic devolatilization history," *Geochim. Cosmochim. Acta* 56 (1992), 2839–2849.

23. D. J. Velinsky, T. C. Hoering, T. Ferdelman, G. Luther III, T. Church, and L. Cifuentes, "Determination of the isotopic composition of elemental sulfur from marine sediment," *CIW Year Book* 89 (1990), 99–100.

24. L. A. Cifuentes, J. H. Sharp, and M. L. Fogel, "Stable carbon and isotope biogeochemistry in the Delaware estuary," *Limnol. Oceanogr.* 33 (1988), 1102–1115.

25. L. A. Cifuentes, M. L. Fogel, J. R. Pennock, and J. H. Sharp, "Biochemical factors that influence the stable nitrogen isotope ratio of dissolved ammonium in the Delaware Estuary," *Geochim. Cosmochim. Acta* 53 (1989), 2713–2721.

26. H. W. Paerl and M. L. Fogel, "Isotopic characterization of atmospheric nitrogen inputs as sources of enhanced primary production in coastal Atlantic Ocean waters," *Marine Bio.* 119 (1994), 635–645.

27. H. W. Paerl, M. L. Fogel, and P. W. Bates, "Atmospheric nitrogen deposition in coastal waters: implications for marine primary production and C flux." In R. Guerrero and C. Pedrós-Alió (eds.), *Trends in Microbial Ecology* (Spanish Society for Microbiology, Barcelona, 1993), pp. 459–464.

28. H. W. Paerl, M. L. Fogel, P. W. Bates, and P. M. O'Donnell, "Is there a link between atmospheric nitrogen deposition and eutrophication in coastal waters?" In K. R. Dyer and R. J. Orth (eds.), *Changes in Fluxes in Estuaries: Implications from Science to Management* (Olsen & Olsen, Fredensborg, Denmark, 1944), pp. 197–202.

29. D. J. Velinsky and M. L. Fogel, "Cycling of dissolved and particulate nitrogen and carbon in the Framvaren Fjord, Norway: Stable isotopic variations," *Marine Chem.* 67 (1999), 161–180.

30. D. J. Velinsky, M. L. Fogel, J. F. Todd, and B. M. Tebo, "Isotopic fractionation of dissolved ammonium at the oxygen–hydrogen sulfide interface in anoxic waters," *Geophys. Res. Lett.* 18 (1991), 649–652.

31. Kerogen is insoluble in water, alkalies, acids and organic solvents, but reacts readily with oxygen. It consists of many very large polymers that can be decomposed at elevated temperatures.

32. P. H. Abelson, "Paleobiochemistry." In *Proceedings of the Fifth International Congress of Biochemistry* (Pergamon Press, 1963), vol. 3, pp. 52–68.

33. Abelson, "Organic geochemistry: kerogen."

34. T. C. Hoering and P. H. Abelson, "Hydrocarbons from kerogen," *CIW Year Book* 62 (1963), 229–234.

35. T. C. Hoering and P. H. Abelson, "Chemicals from the Nonesuch Shale of Michigan," *CIW Year Book* 63 (1964), 262–264.

36. T. C. Hoering, "The extractable organic matter in Precambrian rocks and the problem of contamination," *CIW Year Book* 64 (1965), 215–218.

37. T. C. Hoering and P. H. Abelson, "Fatty acids from the oxidation of kerogen," *CIW Year Book* 64 (1965), 218–223.

38. P. H. Abelson and P. E. Hare, "Uptake of amino acids by kerogen," *CIW Year Book* 68 (1970), 297–303.

39. T. C. Hoering and V. Navale, "A search for molecular fossils in the kerogen of Precambrian sedimentary rocks," *Precam. Res.* 34 (1987), 247–267.

40. G. D. Cody, H. Ade, S. Wirick, G. D. Mitchell, and A. Davis, "Determination of chemical-structural changes in virtrinite accompanying luminescence alteration using C-NEXAFS analysis," *Org. Geochem.* 28 (1998), 441–455.

41. C. D. Cody and G. Sághi Szabó, "Calculation of the ^{13}C NMR chemical shift of ether linkages of lignin derived geopolymers: Constraints on the preservation of lignin primary structure with diagenesis," *Geochim. Cosmochim. Acta* 63 (1999), 193–205.

42. C. K. Boyce, G. D. Cody, M. Feser, C. Jacobsen, A. H. Knoll and S. Wirick, "Organic chemical differentiation within fossil plant cell walls detected with X-ray spectromicroscopy," *Geology* 30 (2002), 1039–1042.

43. C. K. Boyce, G. D. Cody, M. L. Fogel, R. M. Hazen, and C. M. O'D. Alexander, "Chemical evidence for cell wall liquification and the evolution of tracheids in early Devonian plants," *Inter. J. Plant Sci.* (2004) (in press).

44. H. A. Lowenstam, "Biologic problems relating to the composition and diagenesis of sediments." In T. W. Donnelly (ed.), *The Earth Sciences* (University of Chicago Press, 1963), pp. 137–195.

45. O. B. Bøggild, "The shell structure of the mollusks," *K. Dan. Vidensk. Selsk. Skr. Nat. Math.* 2 (1930), 235–325.

46. P. E. Hare and P. H. Abelson, "Amino acid composition of some calcified proteins," *CIW Year Book* 64 (1965), 223–232.

47. Some thirty-nine organic compounds are recognized and approved as minerals by the International Mineralogical Association. Among those is abelsonite, $NiC_{13} H_{32} N_4$, a nickel porphyrin, named after Dr. Philip H. Abelson, past president of CIW. See C. Milton, E. J. Dwornik, P. A. Estep-Barnes, R. G. Finkelman, A. Pabst, and S. Palmer, "Abelsonite, nickel porphyrin, a new mineral from the Green River Formation, Utah," *Am. Mineral.* 63 (1978), 930–937.

48. T. C. Hoering, "Fichtelite hydrocarbons in fossil wood," *CIW Year Book* 67 (1969), 203–205.

49. D. Rumble and T. C. Hoering, "Carbon isotope geochemistry of graphite vein deposits from New Hampshire U.S.A.," *Geochim. Cosmochim. Acta* 50 (1986), 1239–1247.

50. D. Rumble, J. M. Ferry, T. C. Hoering, and A. J. Boucot, "Fluid flow during metamorphism at the Beaver Brook fossil locality, New Hampshire," *Am. J. Sci.* 282 (1982), 886–919.

51. Rumble and Hoering, "Carbon isotope geochemistry."

52. M. L. F. Estep and H. Dabrowski, "Tracing foodwebs with stable hydrogen isotopes," *Science* 209 (1980), 1537–1538.

53. S. A. Macko, M. L. F. Estep, and W. Y. Lee, "Stable hydrogen isotope analysis of foodwebs on laboratory and field populations of marine amphipods," *J. Exp. Mar. Biol. Ecol.* 72 (1983), 243–249.

54. N. Tuross, M. L. Fogel, and P. E. Hare, "Variability in the preservation of the isotopic composition of collagen from fossil bone," *Geochim. Cosmochim. Acta* 52 (1988), 929–935.

55. P. L. Koch, A. K. Behrensmeyer, A. W. Stott, N. Tuross, R. P. Evershed, and M. L. Fogel, "The effects of weathering on the stable isotope composition of bones," *Ancient Biomolecules* 3 (2001), 117–134.

56. A. S. Brooks, P. E. Hare, J. E. Kokis, G. H. Miller, R. D. Ernst, and F. Wendorf, "Dating Pleistocene archaeological sites by protein diagenesis in ostrich eggshell," *Science* 248 (1990), 60–64.

57. Hare, Fogel, Stafford, Mitchell, and Hoering, "The isotopic composition of carbon."

58. P. L. Koch, J. Heisinger, C. Moss, R. W. Carlson, M. L. Fogel, and A. K. Behrensmeyer, "Isotopic tracking of change in diet and habitat use in African elephants," *Science* 267 (1995), 1340–1343.

59. B. J. Johnson, G. H. Miller, M. L. Fogel, J. W. Magee, M. K. Gagan, and A. R. Chivas, "65,000 years of vegetation change in central Australia and the Australian summer monsoon," *Science* 284 (1999), 1150–1152.

60. H. Bocherens, M. L. Fogel, N. Tuross, and M. Zeder, "Trophic structure and climate information from isotopic signatures in Pleistocene cave fauna of southern England," *J. Arch. Sci.* 22 (1995), 327–340.

61. M. L. Fogel, N. Tuross, B. J. Johnson, and G. H. Miller, "Biogeochemical record of ancient humans," *Org. Geochem.* 27 (1997), 275–287.

62. T. R. Filley, R. A. Blanchet, E. Simpson, and M. L. Fogel, "Nitrogen cycling by wood decomposing soft-rot fungi in the King Midas tomb, Gordion, Turkey," *Proc. Nat. Acad. Sci.* 98 (2001), 13346–13350.

63. M. A. Teece, M. L. Fogel, N. Tuross, E. E. Spanier, and R. M. McCourt, "The Lewis and Clark Herbarium of the Academy of Natural Sciences," *Notulae Naturae* 477 (2002), 1–16.

64. J. B. Corliss, J. A. Baross, and S. E. Hoffman, "An hypothesis concerning the relationship between submarine hot springs and the origin of life on Earth," *Oceanol. Acta.* 4 (Suppl.) (1981), 59–69.

65. Estep, "Carbon and hydrogen isotopic compositions of algae."

66. J. A. Baross, "Isolation and cultivation of hyperthermophilic bacteria from marine and freshwater habitats." In P. F. Kemp, B. F. Shear, E. B. Sherr, and S. J. Cole (eds.), *Handbook of Methods in Aquatic Microbial Ecology* (Lewis Publishers, Boca Raton, FL, 1993), pp. 21–30.

67. A. Sharma, J. H. Scott, G. D. Cody, M. L. Fogel, R. M. Hazen, R. J. Hemley, and W. T. Huntress, "Microbial activity at gigapascal pressures," *Science* 295, 1514–1516.

68. G. D. Cody, N. Z. Boctor, T. R. Filley, R. M. Hazen, J. H. Scott, A. Sharma, and H. S. Yoder, Jr., "Primordial carbonylated iron-sulfur compounds and the synthesis of pyruvate," *Science* 289 (2000), 1337–1340.

12 Geophysics

1. F. E. Wright, "Gravity on the earth and on the moon," *Sci. Monthly* 24 (1927), 448–462; Wright, "The surface features of the moon," *Sci. Monthly* 40 (1935), 101–115.

2. F. A. Vening Meinesz and F. E. Wright, "The gravity measuring cruise of the U.S. Submarine S-21 (With an Appendix on computational procedure by Eleanor A. Lamson)," *Publications of the US Naval Observatory, Second Series* 13 (1930), Appendix I, 94 pp.

3. F. E. Wright and J. L. England, "An improved torsion gravity meter," *Am. J. Sci.* 35-A (1938), 373–383.

4. F. E. Wright, "Gravity-measurements in Guatemala," *Trans. Am. Geophys. Union* 22 (1941), 512–515.

5. F. E. Wright and J. L. England, "Gravity measurements," *CIW Year Book* 40 (1941), 47–48.

6. P. M. Bell and R. F. Roy, "Heat flow and gravity measurements at Ajo, Arizona," *CIW Year Book* 65 (1967), 414–415.

7. J. Johnston and L. H. Adams, "On the measurement of temperature in bore-holes," *Econ. Geol.* 11 (1912), 741–762.

8. J. Johnston, "Note on the temperature in the deep boring at Findlay, Ohio," *Am. J. Sci.* 36 (1913), 131–134.

9. S. P. Clark, Jr., "Heat flow in the Austrian Alps," *J. Geophys. Res.* (1961), 54–63.

10. Bell and Roy, "Heat flow and gravity."

11. L. H. Adams, "Temperatures at moderate depths within the Earth," *J. Wash. Acad. Sci.* 14 (1924), 459–472.

12. A. Holmes, "Earth's thermal history," *Geol. Mag.* 53 (1916), 60–71.

13. H. Jeffreys, *The Earth* (Cambridge University Press, 1924), 278 pp.

14. W. D. Urry, "Significance of radioactivity in geophysics: Thermal history of the earth," *Trans. Am. Geophys. Union* 30 (1949), 171–180.

15. S. P. Clark, Jr., "Heat flow from a differentiated earth," *J. Geophys. Res.* 66 (1961), 1231–1234.

16. F. R. Boyd, "A pyroxene geotherm," *Geochim. Cosmochim. Acta* 37 (1973), 2533–2546.

17. E. D. Williamson, "The effect of strain on heterogeneous equilibration," *Phys. Rev.* 10 (1917), 275–283.

18. F. E. Wright and J. C. Hostetter, "The thermodynamic reversibility of the equilibrium relations between a strained solid and its liquid," *J. Wash. Acad. Sci.* 7 (1917), 405–471.

19. R. W. Goranson, "Flow in stressed solids: an interpretation," *Bull. Geol. Sci. Am.* 51 (1940), 1023–1034.

20. D. Griggs, "Experimental flow of rocks under conditions favoring recrystallization," *Bull. Geol. Soc. Am.* 51 (1940), 1001–1022.

21. E. Hansen, *Strain Facies* (Springer-Verlag, 1968).

22. P. M. Bell and H-k. Mao, "The hypothesis of melting at stress dislocations in the Earth," *CIW Year Book* 71 (1972), 416–418.

23. H. S. Yoder, Jr., "Contemporaneous rhyolite and basalt," *CIW Year Book* 69 (1971), 141–145. Yoder, Jr., "Contemporaneous basaltic and rhyolitic magmas," *Am. Mineral.* 58 (1973), 153–171.

24. C. S. Piggot, "Apparatus to secure core samples from the ocean-bottom," *Bull. Geol. Soc. Am.* 47 (1936), 675–684. Piggot, "Core samples of the ocean bottom and their significance," *Sci. Monthly* 47 (1938), 201–217.

25. A. Hoffmann, S. R. Hart, and P. E. Hare, "Sr^{87}/Sr^{86} ratios of pore fluids from deep-sea cores," *CIW Year Book* 71 (1972), 563–564.

26. P. E. Hare, "Amino acid geochemistry of a sediment core from the Cariaco Trench," *CIW Year Book* 71 (1972), 592–596.

27. L. H. Adams and J. W. Green, "The influence of hydrostatic presure on the critical temperature of magnetization for iron and other materials," *Philos. Mag.* 12 (1931), 361–380.

28. E. Posnjak, "Effect of pressure on the Curie temperature," *CIW Year Book* 40 (1941), 45–46.

29. L. Patrick, "The change of ferromagnetic Curie points with hydrostatic pressure," *Phys. Rev.* 93 (1954), 384–392.

30. Y. A. Timofeev, H-k. Mao, V. V. Struzhkin, and R. J. Hemley, "Inductive method for investigation of ferromagnetic properties of materials under pressure," *Rev. Sci. Instr.* 70 (1999), 4059–4061.

31. D. H. Lindsley, "Investigations in the system $FeO-Fe_2O_3-TiO_2$," *CIW Year Book* 61 (1962), 100–106; P. Wasilewski, D. Virgo, G. C. Ulmer, and F. C. Schwerer, "Magnetic properties of spinels synthesized at $1300°C$ in the $FeCr_2O_4-Fe_3O_4$ solid solutions series," *CIW Year Book* 73 (1974), 327–332.

32. E. Posnjak, "Magnetic and X ray diffraction studies of sulfide minerals," *CIW Year Book* 39 (1940), 42.

33. E. F. Osborn, E. B. Watson, and S. A. Rawson, "Composition of magnetite in subalkaline volcanic rocks," *CIW Year Book* 78 (1979), 475–481; Osborn and S. A. Rawson, "Experimental studies of magnetite in calc-alkaline rocks," *CIW Year Book* 79 (1980), 281–285.

34. S. E. Haggerty and E. Irving, "Mid-Atlantic Ridge near 45°N: Magnetism and magnetic mineralogy," *CIW Year Book* 69 (1971), 259–264.

35. S. P. Clark, Jr., "Optical and electrical properties of silicates," *CIW Year Book* 58 (1959), 187–189.

36. H-k. Mao, "Electrical and optical properties of the olivine series at high pressure," *CIW Year Book* (1973), 552–554.

37. H-k. Mao and P. M. Bell, "The ultrahigh-pressure diamond cell: Design applications for electrical measurements of mineral samples at 1.2 Mbar," *CIW Year Book* 75 (1976), 824–827.

38. T. Murase, I. Kushiro, and T. Fujii, "Electrical conductivity of partially molten peridotite," *CIW Year Book* 76 (1977), 416–419.

39. A. Noyes, *The Electrical Conductivity of Aqueous Solutions* (Carnegie Institution of Washington Publication 63, 1907), 352 pp.

40. L. H. Adams and R. E. Hall, "The effect of pressure on the electrical conductivity of solutions of sodium chloride and of other electrolytes," *J. Phys. Chem.* 35 (1931), 2145–2163.

41. J. D. Frantz and W. L. Marshall, "Electrical conductance studies of $MgCl_2-H_2O$ and $CaCl_2-H_2O$ solutions," *CIW Year Book* 78 (1979), 591–596.

42. L. H. Adams and E. D. Williamson, "The compressibility of minerals and rocks at high pressures," *J. Franklin Inst.* 195 (1923), 475–529.

43. H. E. Tatel, L. H. Adams, and M. A. Tuve, "Studies of the earth's crust using waves from explosions," *Proc. Am. Philos. Soc.* 97 (1953), 658–669.

44. S. P. Clark, Jr. and A. E. Ringwood, "Density distribution and constitution of the mantle," *Rev. Geophys.* 2 (1964), 35–88.

45. E. Ringwood, I. E. MacGregor, and F. R. Boyd, "Petrological constitution of the upper mantle," *CIW Year Book* 63 (1964), 147–152.

46. T. Murase and I. Kushiro, "Compressional wave velocity in partially molten peridotite at high pressures," *CIW Year Book* 78 (1979), 559–562.

47. H. S. Yoder, Jr., "High–low quartz inversion up to 10,000 bars," *Trans. Am. Geophys. Union* 31 (1950), 827–835.

48. Murase and Kushiro, "Compressional wave velocity;" Murase and H. Fukuyama, "Shear wave velocity in partially molten peridotite at high pressures," *CIW Year Book* 79 (1980), 307–310.

49. D. Z. Anderson and H. Spetzler, "Partial melting and the low-velocity zone," *Phys. Earth Planet. Inter.* 4 (1970), 62–64.

13 Extraterrestrial petrology

1. T. C. Chamberlin, "Fundamental problems of geology," *CIW Year Book* 4 (1905), 171–185.

2. J. N. Lockyer, "Meteorites," *Proc. Roy. Soc. London* 43 (1887), 117–156.

3. G. H. Darwin, "On the mechanical conditions of a swarm of meteorites, and on theories of cosmogony," *Phil. Trans. Roy. Soc. London* 180 (1889), 1–69.

4. L. H. Adams and H. S. Washington, "The distribution of iron in meteorites and in the earth," *J. Wash. Acad. Sciences* 14 (1924), 333–340.

5. *Ibid.*, p. 340.

6. S. P. Clark, Jr. and G. Kullerud, "Iron meteorites," *CIW Year Book* 59 (1960), 141–144.

7. P. Ramdohr and G. Kullerud, "Meteorites," *CIW Year Book* 60 (1961), 182–183.

8. A. El Goresy and G. Kullerud, "The Cr-S and Fe-Cr-S systems," *CIW Year Book* 67 (1969), 182–187.

9. H. S. Yoder, Jr. and G. Kullerud. "Kosmochlor and the chromite-plagioclase association," *CIW Year Book* 69 (1971), 155–157.

10. W. B. Bryan and G. Kullerud, "Meteorites," *CIW Year Book* 69 (1971), 245–249.

11. D. Virgo, "Recent Fe Mössbauer spectroscopic studies of some common rock-forming silicates," *CIW Year Book* 70 (1971), 215–221.

12. M. G. Seitz, "Spallation recoil tracks in meteoritic whitlockite," *CIW Year Book* 71 (1972), 553–557.

13. M. G. Seitz, D. S. Burnett, and P. M. Bell. "U, Th, Pu fractionation in geologic systems: Early Pu/U abundance in meteorites," *CIW Year Book* 73 (1974), 451–457.

14. I. Kushiro and M. G. Seitz, "Experimental studies on the Allende chondrite and the early evolution of terrestrial planets," *CIW Year Book* 73 (1974), 448–451.

15. M. G. Seitz, "Volatilization of iron-bearing silicates in the presence of carbon," *CIW Year Book* 72 (1973), 666–668.

16. H-k. Mao and P. M. Bell, "Crystal-field effects of trivalent titanium in fassaite from the Pueblo de Allende meteorite," *CIW Year Book* 73 (1974), 488–492.

17. J. R. Cronin, "Amino acids and their derivatives in carbonaceous chondrites," *CIW Year Book* 74 (1975), 617–619.

18. G. D. Cody, C. M. O'D. Alexander, and F. Tera, "Solid-state (^{1}H and ^{13}C) nuclear magnetic resource spectroscopy of insoluble organic residue in the Murchison meteorite: A self-consistent quantitative analysis," *Geochim. Cosmochim. Acta* 66 (2002), 1851–1865.

19. N. Boctor, P. M. Bell, and H-k. Mao, "Shock metamorphic features in Pampa del Infierno chondrite," *CIW Year Book* 78 (1979), 485–488.

20. H-k. Mao, A. El Goresy, and P. M. Bell, "Evidence of extensive chemical reduction on lunar regolith samples from the Apollo 17 site," *CIW Year Book* 73 (1974), 467–473.

21. S. E. Haggerty and H. O. A. Meyer, "Apollo 12: Opaque oxides," *Earth Planet. Sci. Lett.* 9 (1970), 379–387.

22. D. H. Lindsley and C. W. Burnham, "Pyroxferroite: Stability and x-ray crystallography of synthetic $Ca_{0.15}Fe_{0.85}SiO_3$ pyroxenoid," *Science* 168 (1970), 364–367.

23. S. E. Haggerty, F. R. Boyd, P. M. Bell, L. W. Finger, and W. B. Bryan, "Opaque minerals and olivine in lavas and breccias from Mare Tranquillitatis," *Proc. Apollo 11 Lunar Sci. Conf., Geochim. Cosmochim. Acta Suppl.* 1 (1970), 513–538.

24. L. W. Finger, "Fe/Mg ordering in olivines," *CIW Year Book* 69 (1971), 302–305.

25. F. R. Boyd, "Anatomy of a mantled pigeonite from Oceanus Procellarum," *CIW Year Book* 69 (1971), 216–228.

26. Virgo, "Recent Fe Mössbauer."

27. H-k. Mao and P. M. Bell, "Crystal-field spectra," *CIW Year Book* 70 (1971), 207–215.

28. L. A. Taylor, H-k. Mao, and P. M. Bell, "Rust alteration of the Apollo 16 rocks," *CIW Year Book* 72 (1973), 638–643.

29. Mao, El Goresy, and Bell, "Evidence of extensive chemical reduction."

30. A. Muan and J. F. Schairer, "Melting relations of materials of lunar compositions," *CIW Year Book* 69 (1970), 243–245.

31. F. N. Hodges and I. Kushiro, "Apollo 17 petrology and experimental determination of differentiation sequences in model moon compositions," *Proc. Fifth Lunar Sci. Conf., Geochim. Cosmochim. Acta Suppl.* 5 (1974), 505–520; Kushiro and Hodges, "Differentiation of the model moon," *CIW Year Book* 73 (1974), 454–457.

32. R. Ganapathy and E. Anders, "Bulk composition of the Moon and Earth, estimated from meteorites," *Proc. Fifth Lunar Sci. Conf., Geochim. Cosmochim. Acta Suppl.* 5 (1974), 1181–1206.

33. P. M. Bell and H-k. Mao, "Analysis of type-B lunar symplectites: Garnet composition," *CIW Year Book* 74 (1975), 595–598.

34. Seitz, "Volatilization."

35. B. O. Mysen, D. Virgo, and I. Kushiro. "Experimental studies of condensation processes of silicate materials at low pressures and high temperatures, I. Phase equilibria in the system $CaMgSi_2O_6$-H_2 in the temperature range 1200–1500°C and the pressure range (P_{H_2}) 10^{-6} to 10^{-9} bar," *Earth Planet. Sci. Lett.* 75 (1985), 139–146. Mysen and Kushiro, "Condensation, evaporation, melting, and crystallization in the primative solar nebula: Experimental data in the system MgO-SiO_2-H_2 to 1.0×10^{-9} bar and 1870°C with variable oxygen fugacity," *Am. Mineral.* 73 (1988), 1–19.

36. H-k. Mao, "Static compression of simple molecular systems in the megabar range." In A. Polian, P. Loubeyre, N. Boccara (eds.) *Simple Molecular Systems at Very High Density*, (Plenum Press, 1989).

37. R. H. Hazen, H-k. Mao, L. W. Finger, and R. J. Hemley, "Single-crystal x-ray diffraction of n-H_2 at high pressure," *Phys. Rev. B* 36 (1987), 3944–3947.

14 Astrobiology

1. S. S. Prasad and W. T. Huntress, Jr., "A model for gas phase chemistry in interstellar clouds II," *Astrophys. J.* 239 (1980), 151–160.

2. G. Wächterhäuser, "Groundworks for an evolutionary biochemistry: the iron-sulfur world," *Prog. Biophys. Mol. Biol.* 58 (1992), 85–201.

3. G. D. Cody, N. Z. Boctor, R. M. Hazen, J. A. Brandes, H. T. Morowitz, and H. S. Yoder, Jr., "Geochemical roots of autotrophic carbon fixation: Hydrothermal experiments in the system citric acid, H_2O–($\pm$FeS)–($\pm$NiS)," *Geochim. Cosmochim. Acta* 65 (2001), 3557–3576.

4. G. Wächterhäuser, "Pyrite formation, the first energy source for life: A hypothesis," *Sys. Appl. Microbiol.* 10 (1988), 207–210.

5. G. D. Cody, N. Z. Boctor, T. R. Filley, R. M. Hazen, J. H. Scott, A. Sharma, and H. S. Yoder, Jr., "Primordial carbonylated iron-sulfur compounds and the synthesis of pyruvate," *Science* 289 (2000), 1337–1340.

6. G. D. Cody, N. Z. Boctor, J. A. Brandes, T. R. Filley, R. M. Hazen, and H. S. Yoder, Jr., "Assaying the catalytic potential of transition metal sulfides for prebiotic carbon fixation," *Geochim. Cosmochim. Acta* (2004) (in press).

7. R. M. Hazen, T. R. Filley, and G. A. Goodfriend, "Selective adsorption of L- and D-amino acids on calcite: Implications for biochemical homochirality," *Proc. Nat. Acad. Sci.* 98 (2001), 5487–5490.

8. S. L. Miller, "Production of some organic compounds under possible primitive earth conditions," *J. Am. Chem. Soc.* 77 (1955), 2351–2361.

9. A. I. Oparin, *The Origin of Life on the Earth* (USSR Academy of Sciences, Moscow, 1957).

10. E. T. Peltzer, J. L. Bada, S. Schlesinger, and S. L. Miller, "The chemical conditions on the parent body of the Murchison meteorite: some conclusions based upon amino, hydroxy, and dicarboxylic acids," *Adv. Space Res.* 4 (1984), 69–74.

11. R. Hayatsu and E. Anders, "Organic compounds in meteorites and their origins," *Top. Curr. Chem.* 99 (1981), 3–37.

12. J. A. Brandes, R. M. Hazen, H. S. Yoder, Jr., and G. D. Cody, "Early pre- and post-biotic synthesis of alanine: an alternative to the Strecker." In G. A. Goodfriend *et al.* (eds.), *Perspectives in Amino Acid and Protein Geochemistry* (Oxford University Press, 2000), pp. 41–47.

13. J. A. Brandes, N. Z. Boctor, G. D. Cody, B. A. Cooper, R. M. Hazen, and H. S. Yoder, Jr., "Abiotic nitrogen reduction on the early Earth," *Nature* 395 (1998), 365–367.

14. M. L. Fogel and J. H. Scott, "A new molecular recognition instrument for astro-biological applications." In *NASA Astrobiology Institute General Meeting Abstracts* (2001), pp. 176–177.

15. G. J. F. MacDonald, "Calculation on the thermal history of the Earth," *J. Geophys. Res.* 64 (1959), 1967–2000.

16. G. D. Cody, C. M. O'D. Alexander, and F. Tera, "Solid-state (^1H and ^{13}C) nuclear magnetic resonance spectroscopy of insoluble organic residue in the Murchison meteorite: a self-consistent quantitative analysis," *Geochim. Cosmochim. Acta* 66 (2002), 1851–1865.

17. N. Z. Boctor, C. M. O'D. Alexander, J. Wang, and E. Hauri, "The sources of water in Martian meteorites," *Geochim. Cosmochim. Acta* 67 (2003), 3971–3989.

15 Mineral physics

1. W. A. Bassett, "Mineral physics: atomic global," *EOS Trans. Am. Geophys. Union* 67 (1986), 446–447.

2. F. Birch, "Elasticity and constitution of the earth's interior," *J. Geophys. Res.* 57 (1952), 227–286.

3. C. T. Prewitt, *Report of a Workshop, April 24–26, 1986, on Physics and Chemistry of Earth Materials* (National Research Council, National Academy Press, 1987).

4. F. E. Wright, *The Methods of Petrographic Microscopic Research: Their Relative Accuracy and Range of Application* (Carnegie Institution of Washington Publication 158, 1911).

5. H. E. Merwin and E. S. Larsen, "Mixtures of amorphous sulphur and selenium as immersion media for the determination of high refractive indices with the microscope," *Am. J. Sci.* 34 (1912), 42–47.

6. G. Shen, Y. Fei, U. Hålenius, and Y. Wang, "Optical absorption spectra of (Mg, Fe)SiO$_3$ silicate perovskites," *Phys. Chem. Minerals* 20 (1994), 478–482.

7. Y. Fei, D. Virgo, B. O. Mysen, Y. Wang, and H-k. Mao, "Temperature dependent electron delocalization in (Mg,Fe)SiO$_3$–perovskite," *Am. Mineral.* 79 (1994), 826–837.

8. Raman spectroscopy is based on inelastic scattering of light. The scattered light is typically measured at right angles to the incident beam and passed through a spectrometer whereby the output is detected and recorded as a function of frequency and intensity. In this way, structural information from a crystal can be deduced. J. H. Hibben, "An investigation of intermediate compound formation by means of the Raman effect," *Proc. Nat. Acad. Sci.* 18 (1932), 532–538.

9. S. K. Sharma, "Laser-Raman spectroscopy," *CIW Year Book* 77 (1978), 902–904.

10. H-k. Mao, P. M. Bell, and R. J. Hemley, "Ultrahigh pressures: optical observations and Raman measurement of hydrogen and deuterium to 1.47 Mbar," *Phys. Rev. Lett.* 55 (1985), 99–102.

11. I. Kushiro, H. Satake, and S. Akimoto, "Carbonate-silicate reactions at high pressures and possible presence of dolomite and magnetite in the upper mantle," *Earth Planet. Sci. Lett.* 28 (1975), 116–120.

12. H-k. Mao and R. J. Hemley, "New optical transitions in diamond at ultrahigh pressures," *Nature* 351 (1991), 721–724.

13. R. E. Cohen, "First-principles predictions of elasticity and phase transitions in high pressure SiO$_2$ and geophysical applications." In Y. Syono and M. H. Manghnani (eds.), *High-Pressure Research: Application to Earth and Planetary Sciences* (Terra Scientific Pub. Co., Tokyo, 1992), pp. 425–431. K. J. Kingma., R. E. Cohen, R. J. Hemley, and H-k. Mao, "Transformation of stishovite to a denser phase and lower mantle pressures," *Nature* 374 (1995), 243–245.

14. R. J. Hemley, R. E. Cohen, A. Yeganeh-Haeri, H-k. Mao, and D. J. Weidner, "Raman spectroscopy and lattice dynamics of MgSiO$_3$ perovskite." In A. Navrotsky and D. J. Weidner (eds.), *Perovskite: A Structure of Great Interest to Geophysics and Materials Science* (American Geophysical Union, 1989), pp. 35–44.

15. R. Hazen and C. T. Prewitt, "Effects of temperature and pressure on interatomic distances in oxygen-based minerals," *Am. Mineral.* 62 (1977), 309–315.

16. R. M. Hazen and L. W. Finger, "Crystal structures and compressibilities of pyrope and grossular to 60 kbar," *Am. Mineral.* 63 (1978), 297–303.

17. R. M. Hazen and L. W. Finger, "The crystal structures and compressibilities of layer minerals at high pressure II. Phlogopite and chlorite," *Am. Mineral.* 63 (1978), 293–296.

18. R. M. Hazen and L. W. Finger, "Crystal structure and compressibility of zircon at high pressure," *Am. Mineral.* 64 (1979), 196–201.

19. R. M. Hazen and Z. D. Sharp, "Compressibility of sodalite and scapolite," *Am. Mineral.* 73 (1988), 1120–1122.

20. T. C. McCormick, R. M. Hazen, and R. J. Angel, "Compressibility of omphacite to 60 kbar: role of vacancies," *Am. Mineral.* 74 (1989), 1287–1292.

21. H. Yang, R. T. Downs, L. W. Finger, R. M. Hazen, and C. T. Prewitt, "Compressibility and crystal structure of kyanite, Al_2SiO_5, at high pressure," *Am. Mineral.* 82 (1997), 467–474.

22. H. Yang, R. M. Hazen, L. W. Finger, C. T. Prewitt, and R. T. Downs, "Compressibility and crystal structure of sillimanite, Al_2SiO_5, at high pressure," *Phys. Chem. Minerals* 25 (1997), 39–47.

23. R. L. Ralph, L. W. Finger, R. M. Hazen, and S. Ghose, "Compressibility and crystal structure of andalusite at high pressure," *Am. Mineral.* 69 (1984), 513–519.

24. R. J. Angel, R. M. Hazen, T. C. McCormick, C. T. Prewitt, and J. R. Smyth, "Comparative compressibility of end-member feldspars," *Phys. Chem. Minerals* 15 (1988), 313–318.

25. R. T. Downs, R. M. Hazen, and L. W. Finger, "The high-pressure crystal chemistry of low albite and the origin of the pressure dependency of Al/Si ordering," *Am. Mineral.* 79 (1994), 1042–1052.

26. D. C. Tozer, "The electrical properties of the earth's interior," *Phys. Chem. Earth* 3 (1959), 414–436.

27. H-k. Mao and P. M. Bell, "Electrical conductivity and the red shift of absorption in olivine and spinel at high pressure," *Science* 176 (1972), 403–406; Mao and Bell, "Techniques of electrical conductivity measurement to 300 kbar." In M. H. Manghnani and S. Akimoto (eds.), *High Pressure Research: Applications in Geophysics* (Academic Press, 1977), pp. 493–502.

28. M. I. Eremets, E. A. Gregoryanz, V. V. Struzhkin, H-k. Mao, and R. J. Hemley, "Electrical conductivity of xenon at megabar pressures," *Phys. Rev. Lett.* 85 (2000), 2797–2800.

29. H. Kamerlingh-Onnes, "Disappearance of the electrical resistance of mercury at helium temperatures," *Konink. Akad. Wetensch. Amsterdam Proc.* 14 (1911), 113–115; and R. M. Hazen, *The Breakthrough: The Race for the Superconductor* (Summit Books, New York, 1988).

30. R. M. Hazen, L. W. Finger, R. J. Angel, C. T. Prewitt, N. L. Ross, H.-k. Mao, C. G. Hadidiacos, P. H. Hor, R. L. Meng, and C. W. Chu, "Crystallographic description of phases in the Y-Ba-Cu-O superconductor, *Phys. Rev. B* 35 (1987), 7238–7241; Hazen, C. T. Prewitt, R. J. Angel, N. L. Ross, L. W. Finger, C. G. Hadidiacos, D. R. Veblen, P. J. Heaney, P. H. Hor, R. L. Meng, Y. Y. Sun, Y. Q. Wang, Y. Y. Xue, Z. J. Huang, L. Gao, J. Bechtold, and C. W. Chu, "Superconductivity in the high-Tc Bi-Ca-Sr-Cu-O system: Phase identification," *Phys. Rev. Lett.* 60 (1988), 1174–1177; D. R. Veblen, P. J. Heaney, R. J. Angel, L. W. Finger, R. M. Hazen, C. T. Prewitt, N. L. Ross, C. W. Chu, P. H. Hor, and R. L. Meng, "Crystallography, chemistry, and structural disorder in the new high-Tc Bi-Ca-Sr-Cu-O superconductor," *Nature* 332 (1988), 334–337; R. L. Meng, P. H. Hor, Y. Y. Sun, Z. J. Huang, L. Gao, Y. Y. Xue, Y. Q. Wang, J. Bechtold, C. W. Chu, R. M. Hazen, C. T. Prewitt, R. J. Angel, N. L. Ross, L. W. Finger, and C. G. Hadidiacos, "The 120K-superconducting phase in Bi-Ca-Sr-Cu-O," *Mod. Phys. Lett. B* 2 (1988), 543–549; R. M. Hazen, L. W. Finger, R. J. Angel, C. T. Prewitt, N. L. Ross, C. G. Hadidiacos, P. J. Heaney, D. R. Veblen, Z. Z. Sheng, A. El Ali, and A. M. Hermann, "100 K superconducting phases in

the Tl-Ca-Ba-Cu-O system," *Phys. Rev. Lett.* 60 (1988), 1657–1659; Z. Z. Sheng, A. M. Hermann, D. C. Vier, S. Schultz, S. B. Oseroff, D. J. George, and R. M. Hazen, "Superconductivity in the Tl-Sr-Ca-Cu-O system," *Phys. Rev. B* 38 (1988), 7074–7076; D. E. Morris, J. H. Nickel, J. Y. T. Wei, N. G. Asmar, J. S. Scott, U. M. Scheven, C. T. Hultgren, A. G. Markelz, R. M. Hazen, J. E. Post, P. J. Heaney, and D. R. Veblen, "Eight new high temperature superconductors." *Phys. Rev. B* 39 (1989), 7347–7350; R. M. Hazen, L. W. Finger and D. E. Morris, "Crystal structure of DyBa$_2$Cu$_4$O$_8$: A new 77 K bulk superconductor." *Appl. Phys. Lett.* 54 (1989), 1057–1059; R. M. Hazen, "Crystal structures of high-temperature superconductors," In D. M. Ginsberg (ed.), *Physical Properties of High-Temperature Superconductors II* (World Scientific, 1990), pp. 121–198.

31. L. Gao, Y. Y. Xue, F. Chen, Z. Xiang, R. L. Meng, D. Ramirez, C. W. Chu, J. H. Eggert, and H-k. Mao, "Superconductivity up to 164K in HgBa$_2$Ca$_{m-1}$Cu$_m$O$_{2m+2+\delta}$ ($m = 1$, 2, and 3) under quasihydrostatic pressures," *Phys. Rev. B* 50 (1994), 4260–4263.

32. L. Stixrude and R. E. Cohen, "High-pressure elasticity of iron and anisotropy of Earth's inner core," *Science* 267 (1995), 1972–1975.

33. H-k. Mao, J. Shu, G. Shen, R. J. Hemley, B. Li, and A. K. Singh, "Elasticity and rheology of iron above 220 GPa and the nature of the Earth's inner core," *Nature* 396 (1999), 741–743.

34. H. R. Wenk, S. Matthies, R. J. Hemley, H-k. Mao, and J. Shu, "The plastic deformation of iron at pressure of the Earth's inner core," *Nature* 405 (2000), 1044–1047.

35. G. Steinle-Neumann, L. Stixrude, R. E. Cohen, and O. Gülseren, "Elasticity of iron at the temperature of the Earth's inner core," *Nature* 413 (2001), 57–60.

36. T. S. Duffy, C. Zha, R. T. Downs, H-k. Mao, and R. J. Hemley, "Elasticity of forsterite to 16 GPa and the composition of the upper mantle," *Nature* 378 (1995), 170–173.

37. C.-S. Zha, T. S. Duffy, R. T. Downs, H-k. Mao, and R. J. Hemley, "Sound velocity and elasticity of single-crystal forsterite to 16 GPa," *J. Geophys. Res.* 101 (1996), 17535–17545.

38. C.-S. Zha, T. S. Duffy, R. T. Downs, H-k. Mao, R. J. Hemley, and D. J. Weidner, "Single-crystal elasticity of the α and β of Mg$_2$SiO$_4$ polymorphs at high pressure." In M. H. Manghnani and T. Yagi (eds.), *Properties of Earth and Planetary Materials at High Pressure and Temperature* (American Geophysical Union, 1998), pp. 9–16.

39. Y. Fei, H-k. Mao, J. Shu, and J. Hu, "P-V-T equation of state of magnesiowüstite (Mg$_{0.6}$Fe$_{0.4}$)O," *Phys. Chem. Minerals* 18 (1992), 416–422.

40. R. J. Hemley, H-k. Mao, L. W. Finger, A. P. Jephcoat, R. M. Hazen, and C.-S. Zha, "Equation of state of solid hydrogen and deuterium from single-crystal x-ray diffraction to 26.5 GPa," *Phys. Rev. B* 42 (1990), 6458–70.

41. T. S. Duffy, R. J. Hemley, and H-k. Mao, "Equation of state and shear strength at multimegabar pressures: magnesium oxide to 227 GPa," *Phys. Rev. Lett.* 74 (1995), 1371–1374.

42. J. V. Badding, H-k. Mao, and R. J. Hemley, "High-pressure crystal structure and equation of state of iron hydride: Implications for the Earth's core." In Y. Syono and M. H. Manghanani (eds.), *High-pressure Research: Application to Earth and Planetary Science* (Terra Scientific Publ. Co., Tokyo, 1992), pp. 363–371.

43. R. E. Cohen, "Origin of ferroelectricity in oxide ferroelectrics and the difference in ferroelectric behavior of BaTiO₃ and PbTiO₃," *Nature* 358 (1992), 136–138.

44. G. Sághi-Szabó, R. E. Cohen, and H. Krakauer, "First-principles study of piezo-electricity in PbTiO₃," *Phys. Rev. Lett.* 80 (1998), 4321–4324.

45. H. Fu and R. E. Cohen, "Polarization rotation mechanism for ultrahigh electromechanical response in single crystal piezoelectrics," *Nature* 403 (2000), 281–283.

16 Isotopic geochemistry

1. H. C. Urey, "The thermodynamic properties of isotopic substances," *J. Chem. Soc. London* (1947), 562–581.

2. H. Craig, "The geochemistry of the stable carbon isotopes," *Geochim. Cosmochim. Acta* 3 (1953), 53–92.

3. P. H. Abelson and T. C. Hoering, "The biogeochemistry of the stable isotopes of carbon," *CIW Year Book* 59 (1960), 158–165; Abelson and Hoering, "Carbon isotope fractionation on formation of amino acids by photosynthetic organisms," *Proc. Nat. Acad. Sci.* 47 (1961), 623–632.

4. P. L. Parker, "The biogeochemistry of the stable isotopes of carbon in a marine bay," *Geochim. Cosmochim. Acta* 28, (1964) 1155–1164.

5. M. F. Estep and S. Vigg, "Stable carbon and nitrogen isotope traces of trophic dynamics in natural populations and fisheries of the Labontain Lake System, Nevada," *Canadian J. Fish. Aquatic Sci.* 42 (1985), 1712–1719.

6. M. L. F. Estep, "Carbon and hydrogen isotopic compositions of algae and bacteria from hydrothermal environments, Yellowstone National Park," *Geochim. Cosmochim. Acta* 48 (1984), 591–599.

7. E. F. Duke and D. Rumble, "Textural and isotopic variations in graphite from plutonic rocks, South-central New Hampshire," *Contrib. Mineral. Petrol.* 93 (1986), 409–419.

8. D. Rumble and T. C. Hoering, "Carbon isotope geochemistry of graphite vein deposits from New Hampshire, USA," *Geochim. Cosmochim. Acta* 50 (1986), 1239–1247.

9. D. Rumble, N. H. S. Oliver, J. M. Ferry, and T. C. Hoering, "Carbon and oxygen isotope geochemistry of chlorite-zone rocks of the Waterville limestone, Maine, USA," *Am. Mineral.* 76 (1991), 857–866.

10. P. E. Hare, M. L. Fogel, T. W. Stafford, Jr., A. D. Mitchell, and T. C. Hoering, "The isotopic composition of carbon and nitrogen in individual amino acids isolated from modern and fossil proteins," *J. Arch. Sci.* 18 (1991), 277–292.

11. N. Tuross, M. L. Fogel, and P. E. Hare, "Variability in the preservation of the isotopic composition of collagen from fossil bone," *Geochim. Cosmochim. Acta* 52 (1988), 929–935.

12. R. N. Clayton and T. K. Mayeda, "The use of bromine pentafluoride in the extraction of oxygen from oxides and silicates for isotopic analysis," *Geochim. Cosmochim. Acta* 27 (1963), 43–52.

13. R. N. Clayton, J. R. Goldsmith, K. J. Johnson, and R. C. Newton, "Pressure effect on stable isotope fractionation," *EOS Trans. Am. Geophys. Union* 53 (1972), 555.

14. K. Muehlenbachs and I. Kushiro, "Oxygen isotope exchange and equilibrium of silicates with CO₂ or O₂," *CIW Year Book* 73 (1974), 232–236.

15. D. Rumble, "Mineralogy, petrology, and oxygen isotopic geochemistry of the Clough Formation, Black Mountain, western New Hampshire," *J. Petrol.* 19 (1978), 317–340.

16. C. P. Chamberlain and D. Rumble, "Thermal anomalies in a regional metamorphic terrane: an isotopic study of the role of fluids," *J. Petrol.* 29 (1988), 1215–1232.

17. H. P. Taylor and S. Epstein, "$^{18}O/^{16}O$ ratios in rocks and coexisting minerals of the Skaergaard intrusion East Greenland," *J. Petrol.* 4 (1963), 51–74.

18. H. P. Taylor, Jr., "Water/rock interactions and the origin of H_2O in granitic batholiths," *J. Geol. Soc. London* 133 (1977), 177–210.

19. D. E. James, "The combined use of oxygen and radiogenic isotopes as indicators of crustal contamination," *Ann. Rev. Earth Planet. Sci.* 9 (1981), 311–344.

20. D. Rumble, J. M. Ferry, and T. C. Hoering, "Oxygen isotope geochemistry of hydrothermally-altered synmetamorphic granite rocks from South-Central Maine, USA," *Contrib. Mineral. Petrol.* 93 (1986), 420–428.

21. T.-F. Yui, D. Rumble, and C.-H. Lo, "Unusually low $\delta^{18}O$ ultra-high pressure metamorphic rocks from the Sulu Terrain, eastern China," *Geochim. Cosmochim. Acta* 59 (1995), 2859–2864.

22. R. Y. Zhang, D. Rumble, J. G. Liou, and Q. C. Wang, "Low $\delta^{18}O$, ultrahigh-P garnet-bearing mafic and ultramafic rocks from Dabie Shan, China," *Chem. Geol.* 150 (1998), 161–170.

23. Z. D. Sharp, "A laser-based microanalytical method for the *in-situ* determination of oxygen isotope ratios of silicates and oxides," *Geochim. Cosmochim. Acta* 54 (1990), 1353–1357. The laser fluorination of silicates has been reviewed by D. Rumble and Z. D. Sharp, "Laser microanalysis of silicates for $^{18}O/^{17}O/^{16}O$ and of carbonates for $^{18}O/^{16}O$ and $^{13}C/^{12}C$ ratios," *Rev. Econ. Geol.* 7 (1998), 99–119.

24. M. Dole, "The relative atomic weight of oxygen in water and in air," *J. Am. Chem. Soc.* 57 (1935), 2731.

25. R. D. Guy, M. L. Fogel, J. H. Berry, and T. C. Hoering, "Isotope fractionation during oxygen production and consumption by plants." In J. Biggens (ed.), *Progress in Photosynthesis Research*, vol. 3 (M. Nijhoff Publishers, 1987), pp. 597–600.

26. R. D. Guy, M. L. Fogel, and J. A. Berry, "Photosynthetic fractionation of the stable isotopes of oxygen and carbon," *Plant Physiol.* 101 (1993), 37–47.

27. K. W. Mandernack, M. L. Fogel, B. M. Tebo, and A. Usui, "Oxygen isotope analyses of chemically and microbially produced manganese oxides and manganates," *Geochim. Cosmochim. Acta* 59 (1995), 4409–4425.

28. M. L. F. Estep and T. C. Hoering, "Biogeochemistry of the stable hydrogen isotopes," *Geochim. Cosmochim. Acta* 44 (1980), 1197–1206.

29. M. L. F. Estep and H. Dabrowski, "Tracing food webs with stable hydrogen isotopes," *Science* 209 (1980), 1537–1538.

30. S. A. Macko, M. L. F. Estep, and W. Y Lee, "Stable hydrogen isotope analysis of food webs on laboratory and field populations of marine amphipods," *J. Exp. Mar. Biol. Ecol.* 72 (1983), 243–249.

31. M. L. F. Estep and T. C. Hoering, "Stable hydrogen isotope fractionations during autotrophic and mixotrophic growth of microalgae," *Plant Physiol.* 67 (1981), 474–477.

32. B. R. Doe, "Relationships of lead isotopes among granites, pegmatites, and sulfide ores near Balmat, New York," *J. Geophys. Res.* 67 (1962), 2895–2906.

33. H. Puchelt and G. Kullerud, "Sulfur isotope fractionation in the Pb-S system," *Earth Planet. Sci. Lett.* 7 (1970), 301–306.

34. T. C. Hoering, "The isotopic composition of bedded barites from the Archean of Southern India," *J. Geol. Soc. India* 34 (1989), 461–466.

35. H. Puchelt, B. R. Sables, and T. C. Hoering, "Preparation of sulfur hexafluoride for isotope geochemical analysis," *Geochim. Cosmochim. Acta* 35 (1971), 625–628.

36. R. P. Ilchik and D. Rumble, "Sulfur, carbon, and oxygen isotope geochemistry of pyrite and calcite from veins and sediments sampled by borehole CCM-2, Creede Caldera, Colorado," *Geol. Soc. Am. Spec. Paper* 346 (2000), 287–300; D. Rumble and T. C. Hoering, "Analysis of oxygen and sulfur isotope ratios in oxide and sulfide minerals by spot heating with a carbon dioxide laser in a fluorine atmosphere," *Acc. Chem. Res.* 27 (1994), 237–241.

37. N. H. S. Oliver, T. C. Hoering, T. W. Johnson, D. Rumble, and W. C. Shanks, "Sulfur isotopic disequilibrium and fluid–rock interaction during metamorphism of sulfidic black shales from Waterville–Augusta area, Maine, USA," *Geochim. Cosmochim. Acta* 56 (1992), 4257–4265.

38. T. C. Hoering, "Variations in nitrogen-15 abundance in naturally occurring substances," *Science* 122 (1955), 1233–1234.

39. S. A. Macko, "Source of organic nitrogen in mid-Atlantic coastal bays and continental shelf sediments of the United States: isotopic evidence," *CIW Year Book* 82 (1985), 390–394.

40. D. J. Velinsky, J. R. Pennock, J. H. Sharp, L. A. Cifuentes, and M. L. Fogel, "Determination of the isotopic composition of ammonium-nitrogen at the natural abundance level from estuarine waters," *Marine Chem.* 26 (1989), 357–361; L. A. Cifuentes, J. H. Sharp, and M. L. Fogel, "Stable carbon and nitrogen isotope biogeochemistry in the Delaware estuary," *Limnol. Oceanogr.* 33 (1988), 1102–1115; L. A. Cifuentes, M. L. Fogel, J. R. Pennock, and J. H. Sharp, "Biogeochemical factors that influence the stable nitrogen isotope ratio of dissolved ammonium in the Delaware Estuary," *Geochim. Cosmochim. Acta* 53 (1989), 2713–2721.

41. Macko, "Source of organic nitrogen."

42. S. A. Macko and M. L. F. Estep, "Microbial alteration of stable nitrogen and carbon isotopic compositions of organic matter," *CIW Year Book* 82 (1983), 394–398.

43. S. A. Macko and M. L. F. Estep, "Microbial alteration of stable nitrogen and carbon isotopic compositions of organic matter," *Org. Geochem.* 6 (1984), 787–790.

44. P. E. Hare and M. L. F. Estep, "Carbon and nitrogen isotopic composition of amino acids in modern and fossil collagens," *CIW Year Book* 82 (1983), 410–414.

45. T. C. Hoering, "Thermal reactions of kerogen with added water, heavy water, and pure organic substances," *CIW Year Book* 81 (1982), 397–402.

46. G. E. Bebout and M. L. Fogel, "Nitrogen-isotope compositions of metasedimentary rocks in the Catalina Schist, California: Implications for metamorphic devolatilization history," *Geochim. Cosmochim. Acta* 56 (1992), 2839–2849.

47. M. L. F. Estep and S. A. Macko, "Nitrogen isotope biogeochemistry of thermal springs," *Org. Geochem.* 6 (1984), 779–785.

17 Geochronology

1. A. H. Becquerel, "Sur les radiations invisibles émises par les sels d'uranium," *Comptes Rendus Acad. Sci.* 122 (1896), 689–694.

2. E. Rutherford and F. Soddy, "The radioactivity of uranium," *Phil. Mag.* 5 (1903), 441–445.
3. E. Rutherford, "Age of radioactive minerals" [Silliman Lectures, Yale University, March, 1905]. In *Radioactive Transformations* (Charles Scribner's Sons, New York, 1906), pp. 187–191.
4. B. B. Boltwood, "On the ultimate disintegration products of the radio-active elements, Part II. The disintegration products of uranium," *Am. J. Sci., 4th series* 23 (1907), 77–88.
5. C. N. Fenner, "The analytical determination of uranium, thorium, and lead as a basis for age-calculation," *Am. J. Sci.* 16 (1928), 369–381.
6. C. N. Fenner and C. S. Piggot, "The mass-spectra of lead from bröggerite," *Nature* 123 (1929), 793–794.
7. C. S. Piggot and W. D. Urry, "The radium content of an ocean-bottom core," *J. Wash. Acad. Sci.* 29 (1939), 405–410.
8. L. T. Aldrich, G. W. Wetherill, G. L. Davis, and G. R. Tilton, "Radioactive ages of micas from granitic rocks by Rb-Sr and K-Ar methods," *Trans. Am. Geophys. Union* 39 (1958), 1124–1134.
9. G. L. Davis, L. T. Aldrich, G. R. Tilton, G. W. Wetherill, and P. M. Jeffery, "The ages of rocks and minerals," *CIW Year Book* 55 (1956), 161–168.
10. R. G. Tilton, G. L. Davis, G. W. Wetherill, and L. T. Aldrich, "Isotopic ages of zircon from granites and pegmatites," *Trans. Am. Geophys. Union* 38 (1957), 360–371.
11. G. W. Wetherill, "An interpretation of the Rhodesia and Witwatersrand age patterns," *Geochim. Cosmochim. Acta* 9 (1956), 290–292.
12. G. R. Tilton, "Isotope composition of lead from tektites," *Geochim. Cosmochim. Acta* 14 (1958), 323–330.
13. T. E. Krogh and P. M. Hurley, "Strontium isotope variation and whole-rock isochron studies, Grenville Province of Ontario," *J. Geophys. Res.* 73 (1968), 7105–7125; T. E. Krogh and G. L. Davis, "Old isotopic ages in the northwestern Grenville Province, Ontario," *Geol. Assoc. Canada Spec. Paper* 5 (1969), 189–192.
14. T. E. Krogh, "A low-contamination method for hydrothermal decomposition of zircon and extraction of U and Pb for isotopic age determination," *Geochim. Cosmochim. Acta* 37 (1973), 485–494.
15. T. E. Krogh and G. L. Davis, "The production and preparation of ^{205}Pb for use as a tracer for isotope dilution analyses," *CIW Year Book* 74 (1975), 416–417.
16. F. Tera and G. J. Wasserburg, "The evolution and history of mare basalts as inferred from U-Th-Pb systematics." In *Lunar Science VI, Part II, Abstracts* (Lunar Science Institute, Houston, 1975), pp. 807–809.
17. G. L. Davis, "Zircons from the mantle," *CIW Year Book* 77 (1978), 895–897.
18. P. E. Hare, "Amino acid dating – limitations and potential," *Geol. Soc. Am. Abstr. Prog.* 9 (1977), 1004–1005.

18 Element partitioning

1. E. T. Allen and W. P. White, "Diopside and its relations to calcium and magnesium metasilicates," *Am. J. Sci.* 27 (1909), 1–47; N. L. Bowen, "The ternary system: diopside-forsterite-silica," *Am. J. Sci.* 38 (1914), 207–264.
2. L. Atlas, "The polymorphism of $MgSiO_3$ and solid-state equilibria in the system $MgSiO_3$-$CaMgSi_2O_6$," *J. Geol.* 60 (1952), 125–147.

3. H. H. Hess, "Pyroxenes of common mafic magmas," *Am. Mineral.* 26 (1941), 515–535, 573–594.

4. F. R. Boyd and J. F. Schairer, "The system MgSiO₃-CaMgSi₂O₆," *J. Petrol.* 5 (1964), 275–309.

5. B. T. C. Davis and F. R. Boyd, "The join Mg₂Si₂O₆-CaMgSi₂O₆ at 30 kilobars pressure and its application to pyroxenes from kimberlites," *J. Geophys. Res.* 71 (1966), 3567–3576.

6. D. H. Lindsley and S. A. Dixon, "Diopside-enstatite equilibria at 850°C, 5 to 35 kb," *Am. J. Sci.* 276 (1976), 1285–1301.

7. F. R. Boyd, "A pyroxene geotherm," *Geochim. Cosmochim. Acta* 37 (1973), 2533–2546.

8. B. O. Mysen, "Partitioning of iron and magnesium between crystals and partial melts in peridotite upper mantle," *Contr. Mineral. Petrol.* 52 (1975), 69–76.

9. B. O. Mysen and I. Kushiro, "Compositional variations of coexisting phases with degree of melting of peridotite in the upper mantle," *Am. Mineral.* 62 (1977), 843–865.

10. R. H. McCallister, L. W. Finger, and Y. Ohashi, "The equilibrium cation distribution in Ca-rich clinopyroxenes," *CIW Year Book* 74 (1975), 539–542.

11. S. Banno and Y. Matsui, "Eclogite types and partition of Mg, Fe, and Mn between clinopyroxene and garnet," *Proc. Jap. Acad.* 41 (1965), 716–721.

12. B. O. Mysen and K. S. Heier, "Petrogenesis of eclogite in high grade metamorphic gneisses exemplified by the Hareidland Eclogite, Western Norway," *Contrib. Mineral. Petrol.* 36 (1972), 73–94.

13. A. Råheim and D. H. Green, "Experimental determination of the temperature and pressure dependence of the Fe-Mg partition coefficient for coexisting garnet and clinopyroxene," *Contrib. Mineral. Petrol.* 48 (1974), 179–203.

14. A. Finnerty and F. R. Boyd, "Evaluation of thermobarometers for garnet peridotites," *Geochim. Cosmochim. Acta* 48 (1984), 15–27.

15. R. W. Luth, D. Virgo, F. R. Boyd, and B. J. Wood, "Ferric iron in mantle-derived garnets," *Contrib. Mineral. Petrol.* 104 (1990), 56–72.

16. F. R. Boyd, "A pyroxene geotherm," *Geochim. Cosmochim. Acta* 37 (1973), 2533–2546, at p. 2535.

17. F. R. Boyd and J. L. England, "The system enstatite-pyrope," *CIW Year Book* 63 (1964), 157–161.

18. I. D. MacGregor and A. E. Ringwood, "The natural system enstatite-pyrope," *CIW Year Book* 63 (1964), 161–163.

19. D. MacGregor, "The system MgO-Al₂O₃-SiO₂: Solubility of Al₂O₃ in enstatite for spinel and garnet peridotite compositions," *Am. Mineral.* 59 (1974), 110–119.

20. J. Akella, "Garnet pyroxene equilibria in the system CaSiO₃-MgSiO₃-Al₂O₃ and in a natural mineral mixture," *Am. Mineral.* 61 (1976), 589–598.

21. Finnerty and Boyd, "Evaluation of thermobarometers."

22. J. Naldrett and G. Kullerud, "A study of the Strathcona Mine and its bearing on the origin of the nickel-copper ores of the Sudbury District, Ontario," *J. Petrol.* 8 (1967), 453–531.

23. B. O. Mysen, "Rare earth partitioning between crystals and liquid in the upper mantle," *CIW Year Book* 75 (1976), 656–659.

24. B. O. Mysen, "Experimental determination of nickel partition coefficients between liquid, pargasite, and garnet peridotite minerals and concentration limits of

behavior according to Henry's law at high pressure and temperature," *Am. J. Sci.* 278 (1978), 217–243.

25. B. O. Mysen, "Nickel partitioning between olivine and silicate melt: Henry's Law revisited," *Am. Mineral.* 64 (1979), 1107–1114.

26. B. O. Mysen and I. Kushiro, "Pressure dependence of nickel partitioning between forsterite and aluminous silicate melts," *Earth. Planet. Sci. Lett.* 42 (1979), 383–388.

27. B. O. Mysen and D. Virgo, "Influence of melt structure on crystal-liquid trace-element partitioning," *CIW Year Book* 79 (1980), 326–327.

28. H. P. Eugster, "Distribution coefficients of trace elements," *CIW Year Book* 53 (1954), 102–104.

29. H. P. Eugster, "The cesium-potassium equilibrium in the system sanidine-water," *CIW Year Book* 54 (1955), 112–113.

30. M. G. Seitz, "Fractionation of a rare earth element between diopside and a basalt melt at 20 kbar pressure," *CIW Year Book* 73 (1974), 547–551.

31. A. Maseda and I. Kushiro, "Experimental determination of partition coefficients of ten rare earth elements and barium between clinopyroxene and liquid in silicate system at 20 kilobar pressure," *Contrib. Mineral. Petrol.* 26 (1970), 42–49.

32. B. O. Mysen, "Partitioning of samarium and nickel between olivine, orthopyroxene, and liquid: Preliminary data at 20 kbar and 1025°C," *Earth Planet. Sci. Lett.* 31 (1976), 1–7.

33. B. J. Wood, "Samarium distribution between garnet and liquid at high pressure," *CIW Year Book* 75 (1976), 659–662.

34. W. S. Harrison, "Partitioning of REE between minerals and coexisting melts during partial melting of a garnet lherzolite," *Am. Mineral.* 66 (1981), 242–259.

35. B. O. Mysen, "Trace-element partitioning between garnet peridotite minerals and water-rich vapor: Experimental data from 5 to 30 kbar," *Am. Mineral.* 64 (1979), 274–287.

36. B. O. Mysen, "Rare earth element partitioning between (H_2O+CO_2) vapor and upper mantle minerals: experimental data bearing on the conditions of formation of alkali basalt and kimberlite," *Neues Jahrb. Mineral. Abh.* 146 (1983), 41–65.

37. J. Brenan, "Partitioning of fluorine and chlorine between apatite and non-silicate fluids at high pressure and temperature." In *Annual Report of the Director, Geophysical Laboratory, 1990–1991* (1991), pp. 61–67.

38. R. F. Wendlandt and W. J. Harrison, "Rare earth partitioning between immiscible carbonate and silicate liquids and CO_2 vapor," *Contrib. Mineral. Petrol.* 69 (1979), 409–419.

19 Petrofabrics and statistical petrology

1. R. S. Woodward, C. Barus, T. C. Chamberlin, A. A. Michelson, C. R. Van Hise, and C. S. Walcott, "Report of Advisory Committee on Geophysics," *CIW Year Book* 1 (1903), 26–43.

2. G. F. Becker, "Finite homogeneous strain, flow and rupture of rocks," *Bull. Geol. Soc. Am.* 4 (1893), 30–35.

3. F. E. Wright, "Schistosity by crystallization: a qualitative proof," *Am. J. Sci.* 22 (1906), 224–230.

4. T. F. W. Barth, "Structural and petrologic studies in Dutchess County, New York," *Bull. Geol. Soc. Am.* 47 (1936), 775–850.

5. B. Sander, *Gefügekunde der Gesteine* (Julius Springer, Vienna, 1930); W. Schmidt, *Tektonik und Verformungslehre*, (Gebrüder Borntraeger, Berlin, 1932), 208 pp.

6. E. Ingerson, "Fabric analysis of a coarsely crystalline polymetamorphic tectonite," *Am. J. Sci.* 31 (1936), 161–187.

7. E. B. Knopf and E. Ingerson, "Structural petrology, Part II. Laboratory technique of petrofabric analysis," *Geol. Soc. Am. Memoir* 6 (1938), 209–262.

8. S. J. Shand, "A recording micrometer for rock analysis," *J. Geol.* 24 (1916), 394–403.

9. F. Chayes, "A simple point counter for thin-section analysis," *Am. Mineral.* 34 (1949), 1–11.

10. F. Chayes, *Petrographic Modal Analysis* (John Wiley & Sons, New York, 1956).

11. F. Chayes, "Notes on the staining of potash feldspar with sodium cobaltinitrite in thin section," *Am. Mineral.* 37 (1952), 337–340; A. Gabriel and E. P. Cox, "A staining method for quantitative determination of certain rock minerals," *Am. Mineral.* 14 (1929), 290–292; M. L. Keith, "Selective staining to facilitate Rosiwal analysis," *Am. Mineral.* 24 (1939), 561–565.

12. F. Chayes, "The chemical composition of Cenozoic andesite." In A. R. McBirney (ed.), *Proceedings of the Andesite Conference* (Oregon Dept. Geol. Mineral Ind. Bull. 65, 1969), pp. 1–11.

13. F. Chayes, "Silica saturation in cenozoic basalt," *Phil. Trans. Roy. Soc. London* A 271, (1972), 285–296; H. S. Washington, "Chemical analyses of igneous rock," *US Geol. Surv. Prof. Paper* 99 (1917), 1201 pp.

14. H. W. Fairbairn (ed.), "A cooperative investigation of precision and accuracy in chemical, spectrochemical and modal analysis of silicate rocks," *US Geol. Survey Bull.* 980 (1951), 71 pp; F. Chayes, "Another last look at Gl-Wl," *J. Int. Assoc. Math. Geol.* 2 (1970), 207–209.

15. A. Harker, *The Natural History of Igneous Rocks* (Methuen & Co., London, 1909), 384 pp.

16. N. L. Bowen, *The Evolution of the Igneous Rocks* (Princeton University Press, 1928), 334 pp.

17. C. N. Fenner, "The Katmai magmatic province," *J. Geol.* 34 (1926), 673–772.

18. F. Chayes, "Variance-covariance relations in some published Harker diagrams of volcanic suites," *J. Petrol.* 5 (1964), 219–237.

19. F. Chayes, "On correlation in petrography," *J. Geol.* 57 (1949), 239–254; Chayes, "Numerical correlation and petrographic variation," *J. Geol.* 70 (1962), 440–452.

20. F. Chayes and W. S. MacKenzie, "Experimental error in determining certain peak locations and distances between peaks in X-ray (powder) diffractometer patterns," *Am. Mineral.* 42 (1957), 534–547.

21. F. Chayes, "A world data base for igneous petrology," *CIW Year Book* 74 (1975), 549–550. It should be noted that the first electronic publication of petrographic data is a table of Japanese rocks assembled by K. Ono with machine generated norms. K. Ono, *Chemical Composition of Volcanic Rocks in Japan* (Geological Survey of Japan 1962), 441 pp.

22. F. Chayes and J. L. Brandle, "Feasibility study for a publicly accessible rock information system," *CIW Year Book* 73 (1974), 480–488.

23. S. Z. Li and F. Chayes, "A prototype data base for IGCP Project 163-IGBA," *Computers and Geosciences* 9 (1983), 523–526.

20 National defense contributions

1. Bausch & Lamb Optical Co. at Rochester, NY; Bureau of Standards of Pittsburgh, PA; Pittsburgh Plate Glass Co., Charleroi, PA; Spencer Lens Co. at Hamburg, Buffalo, NY; Kueffel & Esser Co. at Hoboken, NJ.

2. F. C. Kracek, N. L. Bowen, and G. W. Morey, "The system potassium metasilicate-silica," *J. Phys. Chem.* 33 (1929), 1857–1879.

3. R. F. Geller, A. S. Creamer, and E. N. Bunting, "The system: $PbO-SiO_2$," *J. Res. Natl. Bur. Standards* 13 (1934), 237–244.

4. G. W. Morey and N. L. Bowen, "The melting relations of the soda-lime-silica glasses," *Trans. Soc. Glass. Tech.* 9 (1925), 226–264.

5. L. H. Adams and E. D. Williamson, "Temperature distribution in solids during heating and cooling," *Phys. Rev.* 14 (1919), 99–114.

6. H. S. Roberts, "The cooling of optical glass melts," *J. Am. Ceram. Soc.* 2 (1919), 543–563.

7. R. B. Sosman, "Some fundamental principles governing the corrosion of a fire clay refractory by a glass," *J. Am. Ceram. Soc.* 8 (1925), 191–204.

8. G. W. Morey, *The Properties of Glass* (American Chemical Society Monograph 77, 1938), 571 pp.

9. Day was indeed a critical consultant to the Corning Glass Works, particularly in regard to establishing a systematic approach to the research on thermally and chemically resistant glasses. A laboratory was set up by Eugene C. Sullivan, a friend and former colleague of Day's. Lantern globes for use along the railroads were made from a new low-expansion lead borosilicate glass that withstood rough handling and the weather. The glass, called Nonex, was unfortunately not resistant to chemicals because of the high boron content. By 1914 Sullivan and his assistant W. C. Taylor had developed a lead-free borosilicate glass that met all the safety specifications for food. The first dish made from the new formula was a pie-plate, so the trade name of the glass became "Py-right." This was modified to "Pyrex" to rhyme with Nonex. Patents were issued to Sullivan and Taylor for the glass (US 1,304,623) and the heating vessel (US 1,304,622) on May 27, 1919. These details are recorded to counter the claim that Pyrex was invented at the Geophysical Laboratory. In the lawsuit of Corning Glass Works vs. Anchor Hocking Glass Corporation (June, 1965), it was noted that the high resistance to breaking of crystallized glasses (known as pyrocerams) was, in fact, first recognized by the French when empty wine bottles were crystallized in a forge. It is alleged that Day went to Corning because post-war Washington was an especially inhospitable place for anyone with German sympathies. Day's wife, Helene Kohlrausch, was the daughter of Friedrich Kohlrausch, head of the German Bureau of Standards, and was not reluctant to express her views. Eventually the marriage was dissolved after Mrs. Day returned to Germany with their three daughters while their son remained as a glass consultant in the USA.

10. E. Posnjak and H. E. Merwin, "Note on the Bucher cyanide process for the fixation of nitrogen," *J. Wash. Acad. Sci.* 9 (1919), 28–30.

11. L. H. Adams and E. D. Williamson, "Some physical constraints of mustard gas," *J. Wash. Acad. Sci.* 9 (1919), 30–35.

12. In addition, with bursts of 100 rounds, the barrel sagged. During World War II the author personally experienced this effect and the hot barrel had to be unscrewed from the machine gun with asbestos gloves and a new replacement, several being

stored by the gun, put in – a very unnerving experience when return fire is received. The life of the newly designed barrel was given as 4000 rounds. It is ironic that in the siege of Carnegie's Homestead steel mills in 1892, the National Guardsmen deployed machine guns against the strikers. It was only the fourth time such Gatling guns were used in the bitter struggle between management and organized labor. See E. F. Osborn, "Machine gun barrel development research of the Geophysical Laboratory, Carnegie Institution of Washington" (unpublished report, 17 November 1945. GL Archives).

13. For example, O. F. Tuttle and W. S. Twenhofel, "Effect of temperature on lineage structure in some synthetic crystals," *Am. Min.* 31 (1946), 569–573.

14. Anonymous, *War Activities of the Trustees and Staff, 1939–1946* (Carnegie Institution of Washington, 1946).

15. L. H. Adams, "Elements of a proposed program for future research" (unpublished report, 3 July 1946. GL Archives).

16. A. Steele and J. Toporski, "Astrobiotechnology." In *Proceedings of the Second European Workshop on Exo/Astrobiology* (European Space Agency Special Publication SP-518, 2002), pp. 235–238.

21 Publications

1. Before 1982–1983, the source of funding was not broken down by the Accounting Office in CIW.

2. In most cases, only researchers at institutions with site licenses to the source journals are able to access the full text of GL publications electronically.

3. About two-thirds of the deceased members of the Geophysical Laboratory have Memorials in the National Academy of Sciences, which may be found in the following numbered volumes:

L. H. Adams	52	S. B. Hendricks	56
E. T. Allen	40	H. H. Hess	43
G. P. Becker	21	J. P. Iddings	69
N. L. Bowen	52	E. S. Larsen	37
A. F. Buddington	57	G. S. Piggot	66
C. W. Cross	32	J. F. Schairer	66
A. L. Day	47	H. S. Washington	60
		F. E. Wright	29

4. S. G. Brush, "The most-cited physical-science publications in the 1945–1954 Science Citation Index, Part 3 – Astronomy and Earth Sciences," *Current Contents* 30, No. 43 (1990), 7–16, at p. 13, table 4.

5. E. Garfield, "The articles most cited in the SCI from 1961 to 1982," *Current Comments* 33 (1985), 3–11.

6. H. S. Yoder, Jr. and C. E. Tilley, "Origin of basalt magmas: An experimental study of natural and synthetic rock systems," *J. Petrol.* 3 (1962), 342–532.

7. "The greatest research on Earth," *Science Watch* 12, no. 6 (November/December 2001), http://www.sciencewatch.com/nov-dec2001/sw_nov-dec2001_page1.htm.

22 Support staff

1. Drilling the 7 $\frac{1}{2}$-inch long hole, $\frac{1}{4}$ inch in diameter, in hard steel for a "rod bomb" (pressure vessel for use up to 850 °C – 5kbar; see Figures 3.6 and 3.7) was particularly difficult and time consuming. After waiting one week for the delivery of a replacement bomb, O. F. Tuttle went to the shop and proceeded to demonstrate that it could be done in a day! By using a newly sharpened drill every few passes and blowing chips out every pass in the lathe, not a drill press, he maintained a straight hole the entire length.

INDEX

Note: page numbers in *italics* refer to figures and tables. Geophysical Laboratory is abbreviated to GL and Carnegie Institute of Washington to CIW in subentries.